Handbook
on
Well Logging

Handbook on Well Logging

by
E. W. "Bill" Sengel

The Institute for Energy Development
Oklahoma City, Oklahoma

Handbook on Well Logging

First Printing: January, 1981

Library of Congress Catalog Card Number: 80-85410

ISBN Number: 0-89419-112-8

Printed at IED Press, Inc., Oklahoma City, Oklahoma, United States of America.

Illustrations by: *Joyce Walton, Walton & Associates, Oklahoma City, Oklahoma*

Editing by: *Peggy J. Durham, The WORD Place Company, Oklahoma City, Oklahoma*

ABOUT THE AUTHOR

E. W. "Bill" Sengel has been in the oil and gas business since 1941 when he began as a field operator for Schlumberger. He is now a consulting Professional Well Log Analyst residing in Bella Vista, Arkansas.

Self-educated, Bill Sengel is proof that success consists of dedication, hard work and daring----along with a natural intellect cultivated by dedication, hard work and daring.

He is a member of the Society of Professional Well Log Analysts and an honorary life member of the Oklahoma City Geological Society. His other writings include work as a contributing author to the IED Handbooks on *Petroleum Exploration* and *Basic Land Management.* An accomplished practical speaker, he teaches the annual IED Institutes on Well Logging, both basic and advanced, and has lectured at several major universities.

ILLUSTRATIONS

TABLE OF CONTENTS

Chapter Four
INTERPRETATION AND QUANTITATIVE ANALYSIS

Chapter Five
SELECTION OF LOGGING PROGRAMS

Chapter Six
OTHER TYPES OF WELL LOGS

Chapter Seven
REFERENCES AND CREDITS

Chapter One

INTRODUCTION

The purpose of this Handbook is to introduce you, the reader, to the world of well logging, especially those logs that generally are considered to be in the family of "Electrical Well Logs". Well logs are continuous recordings versus depth of measurements made in a borehole. These measurements of one or more petrophysical characteristics (described later) of the rocks penetrated by the drill bit are made with specially designed instruments lowered into the hole on a wire line.

There are many types of well logs, other than "Electrical", that are routinely used. Some of these will be described in a later section of this book.

It has been the practice for many years for operators to run one or more logs in practically every well drilled, whether it was a producing well or a dry hole. Consequently, there are many thousands of logs available in every oil and gas province.

The reasons for running logs are many: they are a permanent continuous record of all formations cut by the well bore; they are a source of information desired from the drilling of the hole, such as information used by geologists in their search for oil or gas; they provide information used by production people in testing and completing the well, and, of great importance, they provide data used in the evaluation of producing formations.

It is not the intent or purpose here to make you an expert in the field of well logging, but to expose you to the techniques. There are great quantities of publications available from the service companies and professional organizations (such as the Society of Professional Well Log Analysts, Society of Petroleum Engineers of AIME, American Association of Petroleum Geologists, etc.) that will allow in depth studies to be made. The field is very complex and there should be no attempts to oversimplify it. It must be practiced with full knowledge of the procedures and the effects of variables. It is a case where a little knowledge can be very dangerous, for improper results can be very expensive.

There are many symbols, units, and formulas used in well logging, and you should become well acquainted with them. A list of some is shown in Figures 1 and 1-A. This is only a partial list, but it will suffice until you become more involved in log work. Complete lists are available in books of "Log Interpretation Charts" and other publications easily acquired from the various well logging companies.

In this Handbook theory, instrumentation, and mathematics will be kept to a minimum. There are some basic relationships that are necessary, and they will be introduced at their appropriate places in the text.

§ 1.01 Information Desired From the Wellbore

When an operator causes a hole to be drilled, it may be safe to generally say that he has used all the information available to him to maximize the chances for finding oil or gas. However, the hole must be cut in order to realize the results of his efforts. There are several things the operator hopes to learn as a result of the hole being drilled. Of prime importance is the identification of possible producing formations. Of the many rock layers cut by the bit, only some are porous. The rest may be shales, tight sandstones, hard carbonates, or other types of rock layers. Of the porous zones, only some have proper conditions for hydrocarbon accumulations. They must be in such a position that hydrocarbons migrating from source rocks can get into them, and they must have some sort of trapping mechanism.

Rocks can be identified by their chemical and physical characteristics. These generally are color, texture, hardness, and fossil content. In the case of oil or gas-bearing rocks, there may also be stain and/or odor. These are

LIST OF SYMBOLS

Symbols	*Meaning*	*Customary Units*
a	Electrochemical activity	equivalents per liter
a	Coefficient in F-ϕ relation	
A	Atomic weight	
C	Conductivity	millimhos/meter
C_p, C_{cp}, B_{cp}	Sonic compaction correction factor	$C_p = c\Delta t_{shale}/100$
D	Depth	feet or meters
d	Diameter	inches
E	Electromotive force	millivolts
F	Formation resistivity factor	
G, G(d_i)	Geometrical factor (function of d_i.)	
H, I_H	Hydrogen Index	
h	Thickness (bed, mud-cake, etc.)	feet, meters, inches
I	Index	
J, J(d_i)	Pseudo Geometrical Factor (function of d_i)	
K	SP coefficient	
k	Permeability	millidarcies
M	Quantity used in M-N Plot (see CP-8)	
m	Cementation factor	
N	Quantity used in M-N Plot (see CP-8)	
n	Saturation exponent	
P	Salinity	grams NaCl/gram water or ppm x 10^{-6}
P_c	Capillary pressure	
p	Pressure	pounds per square inch, kilograms per square centimeter, atmospheres
Q_v	Shaliness as given by cation exchange capacity per formation-water volume	milliequivalents/milliliter
q	Dispersed-shale fraction of intermatrix porosity	
R	Resistivity	ohm-meters
r	Radial distance from hole axis	inches
S	Saturation	fraction (or percent) of pore volume
T	Temperature	°F or °C
t	Time	microseconds, seconds, minutes
v	Velocity (Sonic)	feet/second
Z	Atomic number	
α	SP reduction factor	
γ	Specific gravity	(for solids and liquids) density ratio to density of water (for gases) density ratio to density of air
ϕ	Porosity	fraction (of bulk volume), or percent of bulk volume (porosity units)
Δr, L_i	Radial distance from borehole	inches
Δt	Sonic interval transit time	microseconds/foot
$\Delta\phi_{Nex}$	Excavation effect	porosity units, or fraction of bulk volume
λ, I_{an}	Coefficient of anisotropy	
ρ	Density	grams per cubic centimeter
Σ	Neutron capture cross-section	c.u. (capture units) or reciprocal centimeters
τ	Thermal neutron decay time	microseconds

LIST OF SUBSCRIPTS

Subscripts	*Explanation*	*Example*
a, app	Apparent	R_a
abs	Absorption, capture	Σ_{abs}
anh	Anhydrite	
b	Bulk	ρ_b (from Density Log)
clay	Clay	V_{clay}
cor, c	Corrected	Δt_{cor}
cp	Compaction	C_{cp}
dis	Dispersed shale	V_{dis}
dol	Dolomite	Δt_{dol}
e	Electrochemical	E_c
e, eq	Equivalent	R_{we}, R_{mfe}
f, fluid	Fluid	ρ_f
g, gas	Gas	S_g
gxo	Gas in invaded zone	S_{gxo}
gyp	Gypsum	
h	Hole	d_h
h	Hydrocarbon	ρ_h
ha	Apparent value for hydrocarbon	ρ_{ha}
hr	Residual hydrocarbon	S_{hr}
i	Invaded zone (inner boundary)	d_i (see Gen-3)

Continued on next page.

Figure 1

Schlumberger

LIST OF SYMBOLS (Continued)

Subscripts	*Explanation*	*Example*
im, z	Intermatrix (including dispersed shale)	ϕ_{im}
int	Intrinsic value (as opposed to recorded log value)	τ_{int}
irr	Irreducible	$(S_w)_{irr}$
J	Liquid junction	E_J
j	Invaded zone (outer boundary)	d_j (see Gen-3)
k	Electrokinetic	E_k
lam	Laminated shale	V_{lam}
lim	Limiting value	ϕ_{lim}
liq	Liquid	ρ_{liq}
log	Recorded log values	Δt_{log}
ls	Limestone	
m	Mud	R_m
max	Maximum value	ϕ_{max}
ma	Matrix	Δt_{ma}
mc	Mud cake	R_{mc}
mf	Mud filtrate	R_{mf}
mfa	Mud filtrate, apparent	R_{mfa}
min	Minimum value	
ni	Non invaded zone	R_{ni}
o (except when used with resistivity)	Oil	S_o
o (with resistivity only), 0 (zero)	100% water saturated	R_o, R_0
PSP	Pseudo Static SP	E_{PSP}
pri	Primary	ϕ_{pri}
r	Relative	k_{ro}, k_{rw}
r	Residual	S_{or} (residual oil saturation)
s	Adjacent bed	R_s
sd	Sandstone	R_{sd}
sec	Secondary	ϕ_{sec}
sh	Shale	V_{sh}
silt	Silt	I_{silt}
SP	Spontaneous Potential	E_{SP}
SSP	Static SP	E_{SSP}
str	Structural shale	V_{str}
t, ni	True (uncontaminated, non invaded)	R_t
T	Total	C_T
w	Formation water	S_w
wa	Formation water, apparent	R_{wa}
xo	Flushed zone	R_{xo}
z, im	Intermatrix (im preferred)	
0 (zero)	100% water saturated	R_0
IL [also ILd (deep) and ILm (medium)]	From Induction Log	R_{IL}, R_{ILd}, R_{ILm}
LL (also LL3, LL7, LL8, LLd and LLs)	From Laterolog	R_{LL}, R_{LL3}, R_{LL7}, R_{LL8}, R_{LLd}, R_{LLs}
6FF40	From 6FF40 IL	R_{6FF40}
MLL	From Microlaterolog	R_{MLL}
PL	From Proximity Log	R_{PL}
16"	From 16" Normal	$R_{16"}$
64"	From 64" Normal	$R_{64"}$
18'8"	From 18' 8" Lateral	$R_{18'8"}$
1"x1"	From 1"x1" microinverse (ML)	$R_{1"x1"}$
2"	From 2" micronormal (ML)	$R_{2"}$
D	From Density Log	ϕ_D
N, SNP, CNL	From Neutron Log, SNP Neutron CNL Neutron	ϕ_N, ϕ_{SNP}, ϕ_{CNL}
S	From Sonic Log	ϕ_S
ND	From Neutron and Density Logs	ϕ_{ND}
Miscellaneous		
S_{wA}	Water saturation from Archie formula	
S_{wR}	Water saturation from ratio method (with $S_{xo} = S_w^{1/5}$)	

ABBREVIATIONS

BHT	Bottom-hole temperature
FT	Formation temperature
GR	Gamma Ray
HI	Hydrogen Index
SP	Spontaneous Potential
SPI	Secondary Porosity Index
PSP	Pseudo-Static SP
SSP	Static SP

Figure 1-A

Figure 1-A List of Symbols

direct measurements and cores or drill cuttings are necessary for them to be made.

There are also petrophysical characteristics that can help identify the rocks. These are electrical (resistivity and spontaneous potential), elasticity, electron density, radioactivity, nuclear, and thermal. They are, or can be, indirectly measured by using different logging devices lowered into the borehole.

If possible producing zones are located, they must be evaluated. The pores (void spaces) in the rock provide the storage space for the oil or gas, so it is necessary to determine the porosity of the zone to see if there is enough to make a commercial reservoir.

Generally, the rocks are deposited in water so that in the beginning the porosity is filled with water. The hydrocarbons migrate into the porosity later and will displace some of the original water. It is necessary to determine whether enough water has been displaced so that only hydrocarbons will be produced.

Another bit of information needed is to determine whether or not there is enough permeability to allow the zone to produce in sufficient quantities. Permeability is a measure of the ability of rock to allow fluids or gas to flow through it.

Then, of course, thickness of the zone is important. Thickness related to porosity and hydrocarbon saturation helps to determine the economic value of the reservoir.

Correlation of various rocks also is desired. The relative position of particular zones with depth is used in subsurface mapping for exploratory and developmental work.

Accurate depth control is important for completion techniques. Perforations, packers, and bridge plugs must be correctly placed, especially when thin zones or water contacts are involved.

Well logs can provide information and data that help satisfy all these requirements. Well logs are used by many people, from the explorationist who might originate the drilling prospect, to the operator who gets the hole drilled, to the production people who complete the well, to the petroleum engineer who determines the economics, and so on.

From this can be seen the importance of properly selected logs and their ultimate use.

Let us now proceed to discuss the various measurements made with logging equipment and how they are used. The first of these will be resistivity, since it is one of the most important.

§ 1.02 Resistivity

In this section we will discuss what resistivity is and what affects it, as well as what happens to a porous permeable formation when exposed to drilling mud during the drilling of the hole, how these affects are overcome with proper logging techniques, and which resistivity logs are used for the various environments in the bore hole.

Resistivity is the ability of a substance or material to impede the flow of electrical current through itself. If electricity can flow through the substance or material very easily, then the substance has low resistivity. If electricity cannot flow easily, then the substance or material has high resistivity. Conductivity is the reciprocal or opposite of resistivity. In the case of low resistivity the conductivity would be high, and with high resistivity the conductivity would be low.

In well logging the resistivity measurements primarily will be a function of rocks and fluids. The fluids involved are those in the void spaces in the rocks and those in the bore hole.

Resistivity is measured in ohms per unit of volume. The unit of volume in logging is a one meter cube, and is expressed on the log as $OHMS.M^2/M$ or ohm-meters. The range of resistivity generally encountered can vary from a few tenths of an ohm-meter to several thousand ohm-meters.

Generally, sedimentary rocks are made up of various minerals such as quartz, silicates and carbonates, and when they are dry they have high resistivity (or low conductivity). Hydrocarbons are also non-conductors, so they have high resistivity. There are conductive solids in nature also, such as pyrite, magnetite, vermiculite, illite, and glauconite, to name a few. These may be found in formations in sufficient quantity to affect the formation resistivity. They can create problems when analyzing the formation from logs, but fortunately this occurance is not too common.

Since sedimentary rocks were deposited in water the void spaces in the rocks will be filled with the water. Sometimes some of the formation water will be displaced by hydrocarbons, but not completely. Therefore the water contained in the void spaces will have a definite effect on the resistivity of the formation. Truly the resistivity is measured through the water contained in the pore spaces.

Water is a conductor of electricity, therefore it has low resistivity. Its resistivity is inversely proportional to its salinity, at a given temperature. High salinity water has low resistivity, and as the salinity decreases the resistivity increases. Temperature will affect the water resistivity also. For a given salinity the resistivity will decrease as the temperature increases. Water found in most subsurface formations has high salinity; therefore it has low resistivity. Temperature generally increases with depth so the formation waters have decreasing resistivities as the formations get deeper. For example, water with 500 PPM NaCl (sodium chloride) would have resistivity of 10 ohm-meters at 75° F., while water with 140,000 PPM NaCl would have resistivity of .06 ohm-meters at 75° F. Both of these values would decrease as the temperature increases.

Formation water resistivity is very important in working with logs. During the earlier years, formation water resistivities quite often were determined indirectly, and sometimes left a lot to be desired so far as accuracy was concerned. For several years there has been much effort spent in obtaining good values from produced water from many zones. This information has been gathered and put into catalogues, so that now there is very much data available concerning formation water resistivities in most areas. This has been a combined effort by everyone, including operators (both majors and independents), service companies, and professional groups. The importance of good formation water resistivities will be explained and developed further when we get to the section on quantitative analysis of logs.

The fluid in the bore hole also has an important role in well logging. Very often it is the deciding factor in the selection of logs to run in the hole. Since practically all wells are drilled by using the rotary method, the fluid in the bore hole will be the drilling mud. The development of a drilling mud program may require the attention of an engineer specially trained in this field. It can be complex and expensive, and there will be no attempt to delve deeply into the subject here. Drilling mud in a bore hole serves several purposes, among which are: containing formation pressures; cooling and lubricating the drill bit; carrying the drill cuttings out of the hole, and stabilizing the hole by plastering the wall with solids from the mud system. These are done by circulating the mud with special pumps through a closed system. The system is, down the inside of the drill pipe, out through the bit, up the annular space between the drill pipe and the hole wall, through a cleaning system to a pit, and then down the pipe, to repeat the cycle.

Most of the time the drilling mud is made up of water mixed with clays and chemicals. The water may be either fresh or salty, but in any event it is the continuous wetting phase. Sometimes the drilling fluid will be oil, or oil mixed with clays and chemicals. In this case the oil is the continuous wetting phase. There are also cases where the well is drilled by circulating air or gas through the system, and there is no fluid involved. All of these situations can dictate which type of logs should be used in the well bore, because all measurements are made with instruments lowered into the individual environments.

In a water-based mud system the water will conduct electrical current so the mud will have a definite resistivity. The resistivity will range from fractions of an ohm-meter in the case of salty make-up water, to several ohm-meters if the make-up water is fresh. Oil or oil-based muds, and or air or gas systems, will not conduct electrical current so their resistivities will be very high, approaching infinity. Each of these conditions will be discussed in the text at their appropriate places.

§ 1.02(a) Resistivity Measurements

Now is the time to introduce a basic relationship pertaining to formation resistivity. Anyone who grasps an understanding of this formula will be well on the way to understanding resistivity measurements as presented on well logs.

The relationship is:

$$\text{Formation resistivity} = \frac{Rw}{\phi^{m} \times S_w{}^{n}} \qquad (1)$$

Where:

Rw = formation water resistivity
ϕ = formation porosity (% of total rock volume)
m = formation cementation factor
S_w = formation water saturation (% of porosity)
n = saturation exponent (generally 2)

This indicates that the formation resistivity is directly proportional to the resistivity of the formation water, and inversely proportional to the amount of porosity and how much water is in the porosity. If a formation has no porosity there is no space for water, so we have, in effect, dry rocks. The resistivity in this case will be very high. As porosity increases and water saturation is 100 percent the formation resistivity decreases. If a formation has a given porosity and it remains constant, then the resistivity will increase if the water saturation decreases from 100 percent. This is what will happen when some of the formation water is displaced by hydrocarbons moving into the porous space. If we have a given porosity and water saturation and they remain constant, the formation will have low resistivity if the water is salty or high resistivity if the water is fresh.

It is important to realize the formation resistivity is measured through the water contained in the porous spaces, and that the water has a direct effect on the resistivity. However, the rock itself has considerable effect on the resistivity. The resistivity contributed by the rock is called formation resistivity factor, commonly shortened to formation factor. The symbol for this is a capital F. An understanding of formation factor is imperative, so we will take some time to delve into it.

In Figure 2 we have an insulated square container one meter on all sides, open at the top. It is level full with water. The resistivity of this meter cube is measured between two opposite faces and gives a value, for example, equal to 1. Since there is 100 percent porosity and 100 percent water saturation, the resistivity measured will be the resistivity of the water. Since the container is a square cubic meter and the unit of resistivity in well logging is OHMS.M^2/M, we have a water resistivity equal to 1 ohm-meter. It should always have a temperature associated with it also. In this case we could assign room temperature of, say, 70° F. Now in our example we have $R_w = 1$ @ 70° F.

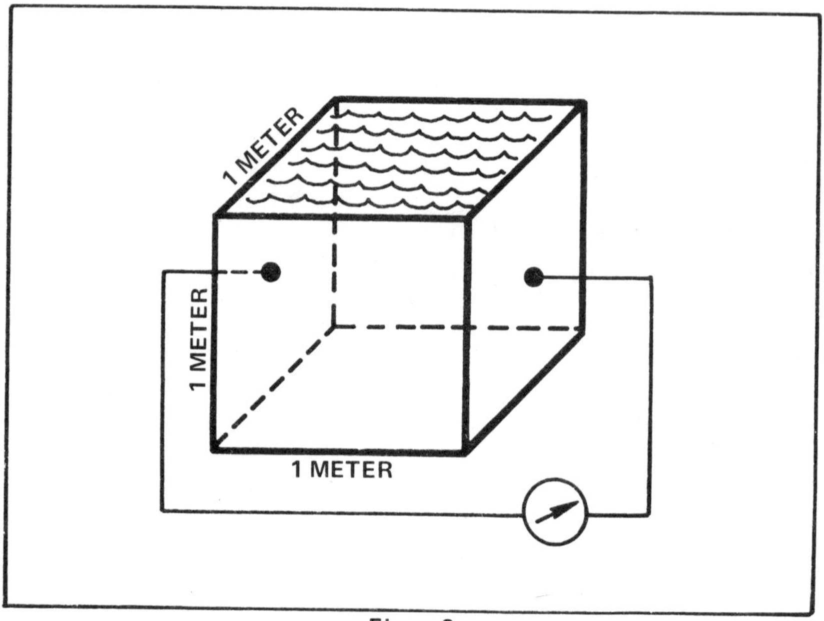

Figure 2

Figure 3 has the same container, and the difference is that sand grains have been added. The water is the same as in the previous figure. The addition of sand has decreased the porosity, however, and the water saturation is still 100%. The resistivity measured between the two opposite faces of the cube will be somewhat higher than before. Let's say that the resistivity measured is now 10 ohm-meters. The additional resistivity is due to the presence of the rock. This is the formation resistivity factor (F). In this case F = 10.

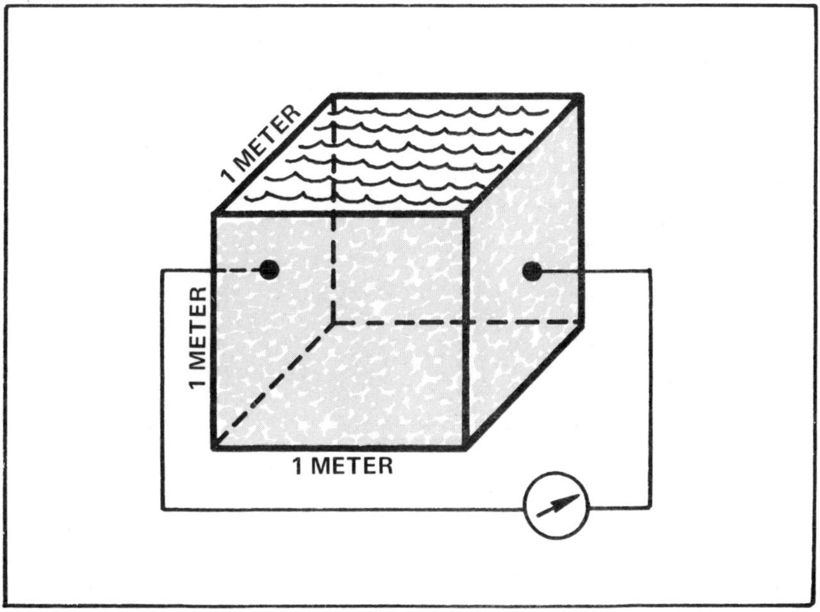

Figure 3

Figure 4 shows the container with the same situation, only now the sand grains have been stuck together with some sort of cementing material. The resistivity measured between the two opposite faces will be higher still, for example, 15 ohm-meters. The water saturation is still 100% and Rw is still 1 ohm-meter. Porosity is the same except now there is cementation involved. Formation factor (F) is now 15. Formation factor can be defined as the number by which water resistivity (Rw) can be multiplied to determine the resistivity of a formation at 100 percent water saturation. The resistivity of a formation at 100 percent water saturation (S_W) is always indicated by the symbol R_O.

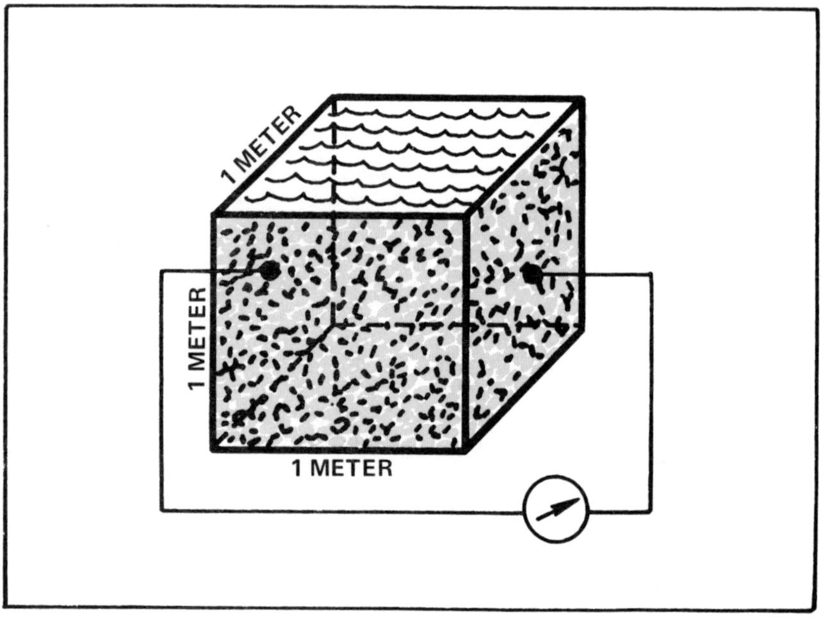

Figure 4

Therefore, $\mathbf{R_o = F \times R_w}$. Mr. G. E. Archie proposed one of the first relationships for F. It is $\mathbf{F} = \frac{1}{\phi^m}$, where the exponent "m" is the cementation factor, and may vary from 1.3 in soft unconsolidated sandstones to as high as 3 in some carbonates. Some knowledge of the rocks is necessary in order to use this formula. Through research a formula was developed to fit most average sandstones. It is the Humble formula and is $\mathbf{F} = \frac{.62}{\phi^{2.15}}$.

In Carbonates $\mathbf{F} = \frac{1}{\phi^2}$ is used, but "m" may vary and increase considerably due to local conditions.

In order to see how F fits into formation resistivity, let's look at formula (1) again. It is: **Formation resistivity** $= \frac{R_w}{\phi^m \times S_w{}^n}$. If we change this to **resistivity** $= \frac{1 \times R_w}{\phi^m \times S_w{}^n}$, and remove $\frac{1}{\phi^m}$ we have our old friend F. So now we can see that: **Formation resistivity** =

$$\frac{F \times R_w}{S_w{}^n} \quad \text{(n generally 2)} \qquad (2).$$

Please review this section on "F" and understand it well, because it will make the rest of the discussion on resistivity much easier to comprehend.

The purpose of resistivity measurements in well logging is to determine the true resistivity of the formation in place, overcoming the effects of the bore hole and invasion by the mud filtrate into the formation.

Let us turn our attention to Figure 5, a schematic presentation of what happens when a porous, permeable formation is penetrated by the drill bit using a water-based mud system.

The porous permeable zone is the shaded interval. As the bit penetrates this zone, filtrate from the drilling mud begins to enter the porous spaces, flushing formation fluids ahead of it. As this action is taking place, solids in

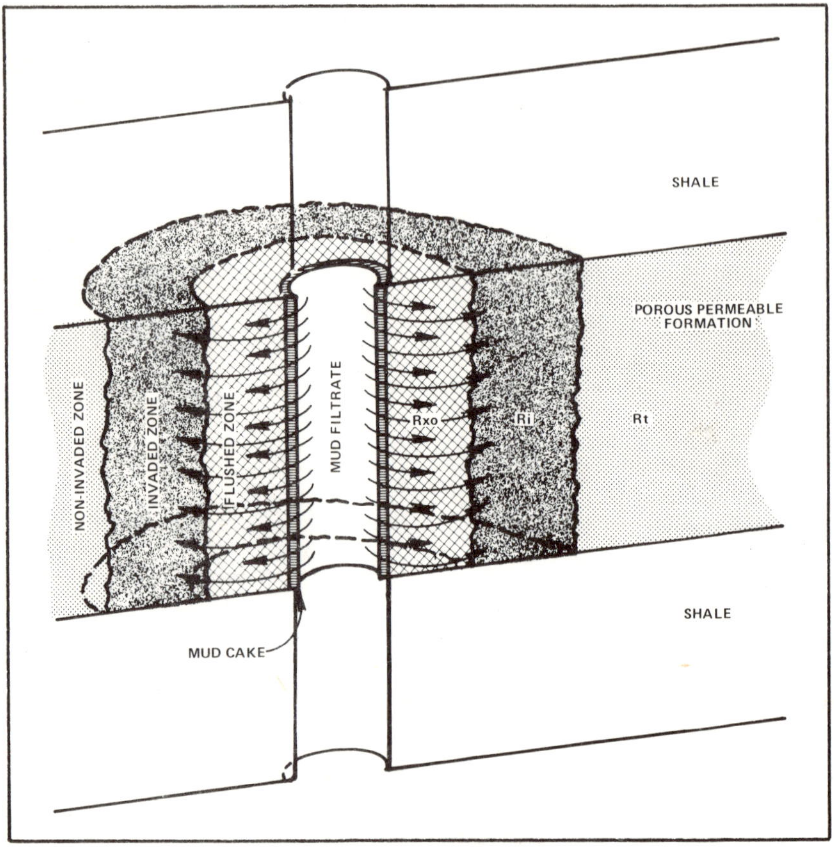

Schematic Illustration of Invasion Process
Figure 5

the mud are plastering the wall of the hole forming mud cake. The mud characteristics control these actions. The action of the filtrate entering the formation is called invasion. The invading fluid will flush out all of the movable formation fluids for some small distance away from the wall of the hole and create the "flushed zone". Invasion may continue, but flushing will become less as it gets farther away from the wall of the hole. At some point there is no more invasion and beyond that point the formation contains only the original formation fluids. The interval between the flushed zone and the non-invaded (or uncontaminated) zone is the "invaded zone". The invaded zone can have many conditions depending on time, mud characteristics, rock characteristics, and original formation fluids, but in any event, it is a problem to be overcome in obtaining true resistivity (R_t). True resistivity is the logging terminology to designate the non-invaded or uncontaminated part of the formation. The fluids in this part are formation water and, hopefully, oil and/or gas. The water saturation in this zone is S_w. The resistivity of the flushed zone is called R_{xo}. The fluid in the flushed zone is primarily mud filtrate (R_{mf}) with some of the original fluids left. The saturation of mud filtrate in the flushed zone is S_{xo}. The resistivity of the invaded zone is called R_i. The fluid in the invaded zone is a mixture of mud filtrate and formation fluids and called R_z. The saturation of fluid mixture in the invaded zone is called S_i.

The adjacent beds are generally shales or impervious formations that are not invaded by the mud filtrate.

Through a series of schematic illustrations we will now look at some invasion profiles that can be expected for various hole conditions.

Figure 6 shows the profile for the situation where there is fresh mud in the hole and the formation is 100 percent saturated with salty water. The fresh mud filtrate invades the formation and displaces the formation water almost completely, creating the flushed zone. The resistivity of this zone equals $F \times R_{mf}$, and will be fairly high because the salty formation water has been replaced by the fresh mud filtrate which has high resistivity. Invasion continues, but flushing is less for some distance creating the invaded zone.

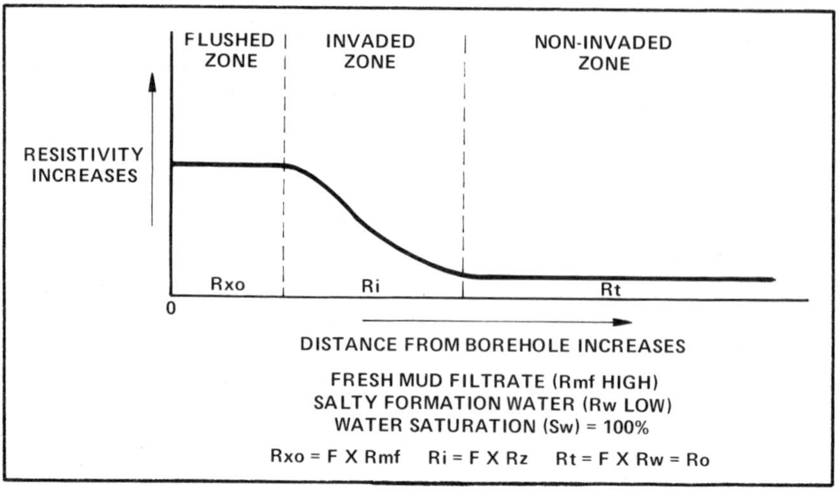

Figure 6

The resistivity of the invaded zone equals F x R_z (mixture of mud filtrate and formation water) and will be high also, because the mixture of filtrate and formation water has higher resistivity than formation water only. Beyond the invaded zone is the non-invaded part of the formation. The resistivity of this zone equals F x R_w and will be low because the salty formation water has low resistivity.

Figure 7 shows the same condition, only now for an oil or gas bearing formation. Each zone of resistivity will have the same relationship as in Figure 6, only now the flushed and invaded zones will be affected by residual oil or gas (that which was not moved out by the invasion process) and the non-invaded zone will be affected by the total oil or gas in the formation. Consequently, as shown in the illustration, each resistivity zone is affected by the reduced water saturation in that zone.

Figure 8 is for the condition where the mud in the hole is salty. Formation water also is salty and S_w is 100%. All zones of resistivity will be equal because salty mud filtrate is displacing salty formation water.

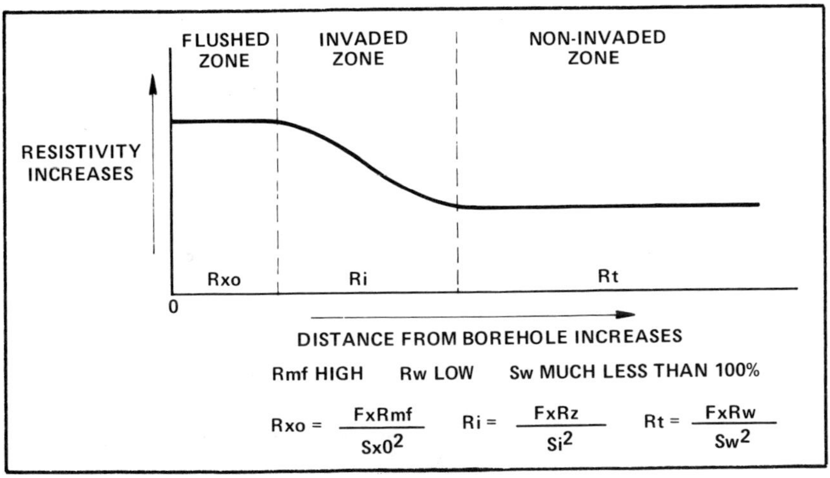

Figure 7

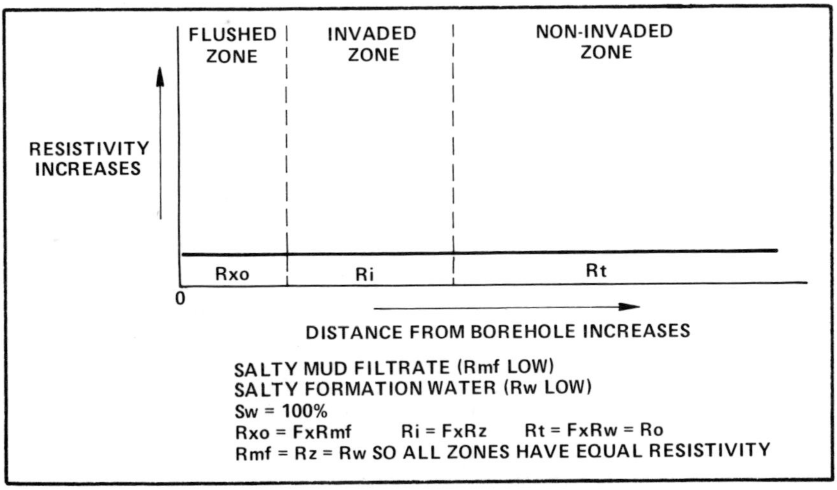

Figure 8

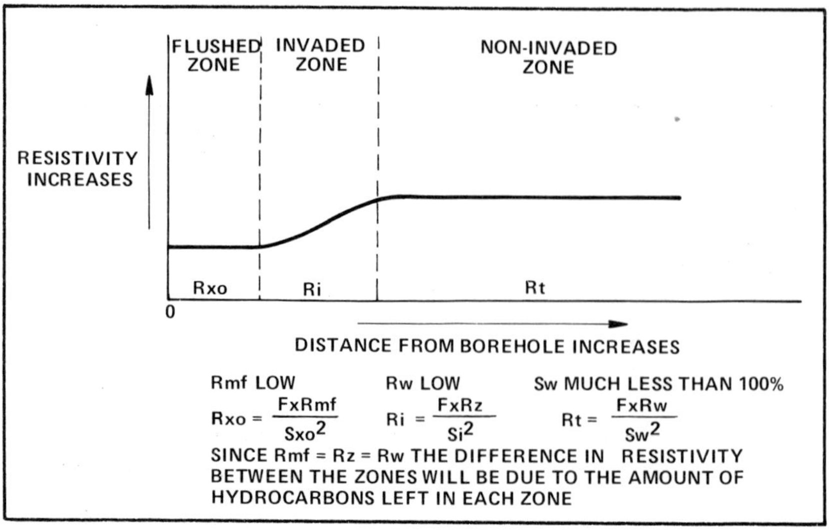

Figure 9

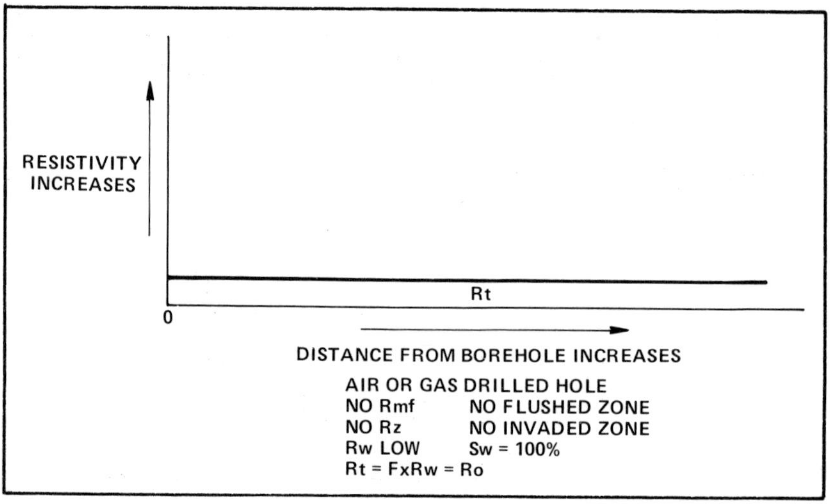

Figure 10

Figure 9 shows an oil or gas bearing formation and salty mud filtrate. Here the invasion by salty filtrate water causes the flushed and invaded zones to have less resistivity than the non-invaded zone.

Figure 10 is where the hole is full of air or gas and the formation is filled with salty water. Since there is no filtrate involved there will be no invasion, and the R_t will be low because the salty formation water has low resistivity. This condition seldom will be seen, though, because water-filled formations may flow, and cause an air or gas-drilling operation to be converted to mud-drilling operation so the water flow can be controlled.

Figure 11 shows an oil or gas-bearing formation with an air or gas-filled hole. Again, there is no invasion because there is no filtrate. R_t will be high because of the reduced S_w.

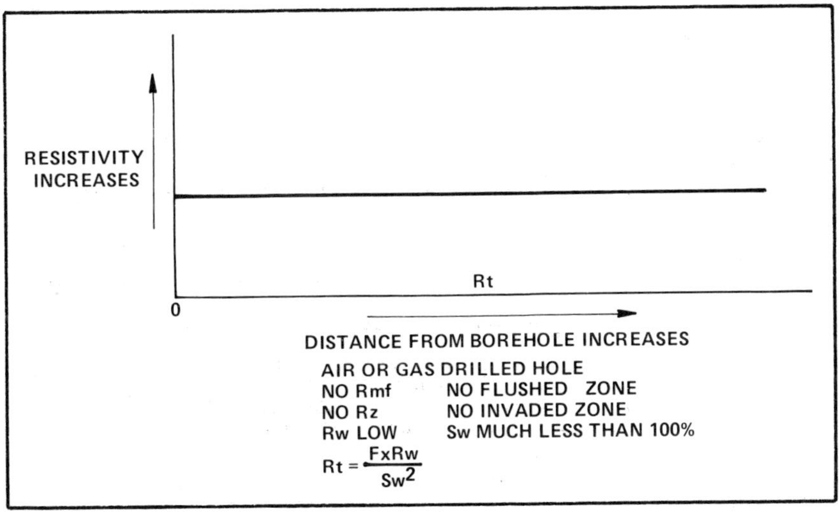

Figure 11

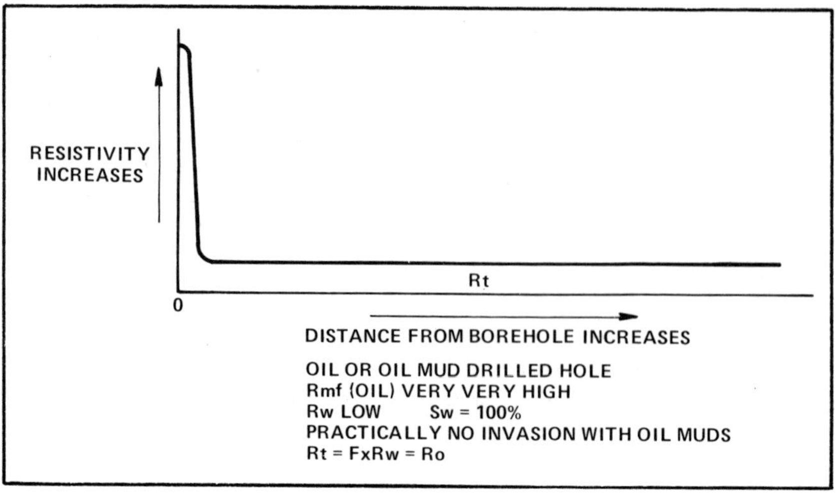

Figure 12

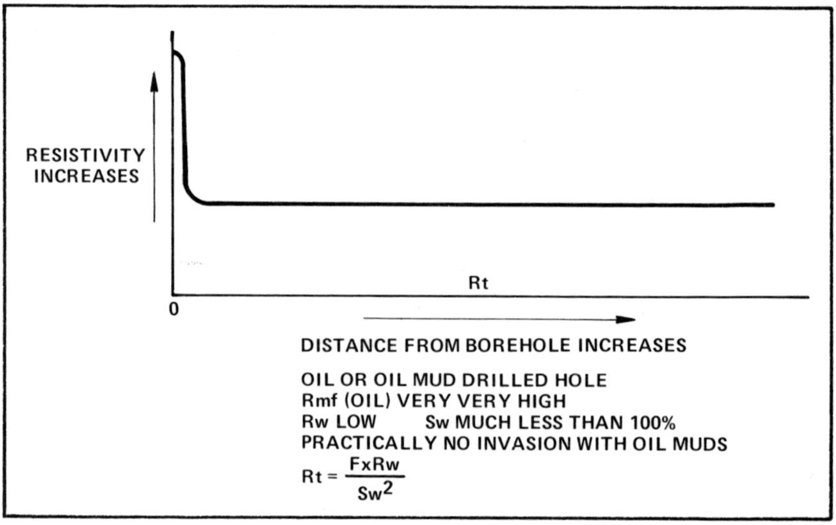

Figure 13

Figures 12 and 13 are for oil or oil mud filled holes. Generally with these hole fluids there is practically no invasion, so only R_t will be measured in both cases.

As can be seen from these illustrations, resistivity measured in a formation will vary as distance from the bore hole wall increases. Resistivity measuring instruments are designed to have particular depths of investigation. They have been designed to measure predominately in the flushed zone, the invaded zone, and, hopefully, in the non-invaded zone. Generally it takes a special instrument to measure only the flushed zone, since it is such a small distance. Instruments that measure in the invaded zone and beyond are designed to make several measurements simultaneously, each progressively further into the formation. The general practice by most all the logging companies is to present three resistivity curves, each with its own depth of investigation. There will be a shallow investigation curve which measures predominantly within the invaded zone; a medium investigation curve which may or may not be affected by the invaded zone (depending on how deep the filtrate invades), and a deep investigating curve which, hopefully, measures beyond the invaded zone and determines true resistivity. The use of each measurement will be discussed at length when we get to the section on quantitative analysis of the logs.

§ 1.02(b) Resistivity Logs

Basically, the resistivity logs are of two types; conventional and focused. Conventional logs are considered to be the "old style" logs and are not run too often now. However, there are many thousands of them in files everywhere, and they are in constant use by explorationists in their search for oil and gas. A knowledge of them is important. Conventional logs had adverse effects caused by the bore hole, formation thickness, formation resistivity, mud resistivity, and invasion. To overcome these effects focused type logs were developed, which allow more accurate analyses to be made and enhance the use of computers in log work.

Each logging company has its own version of each type of log. The

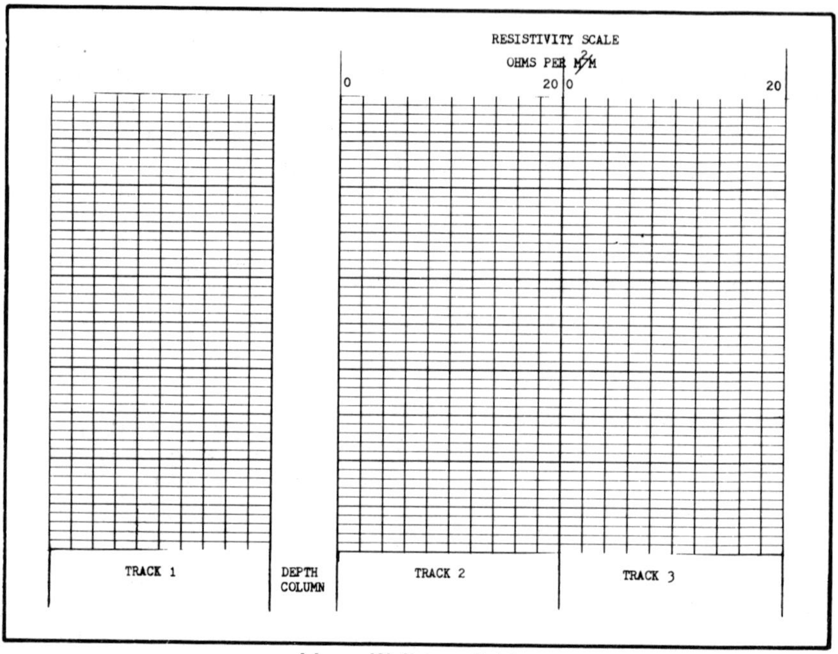

Linear Well Log Grid
Figure 14

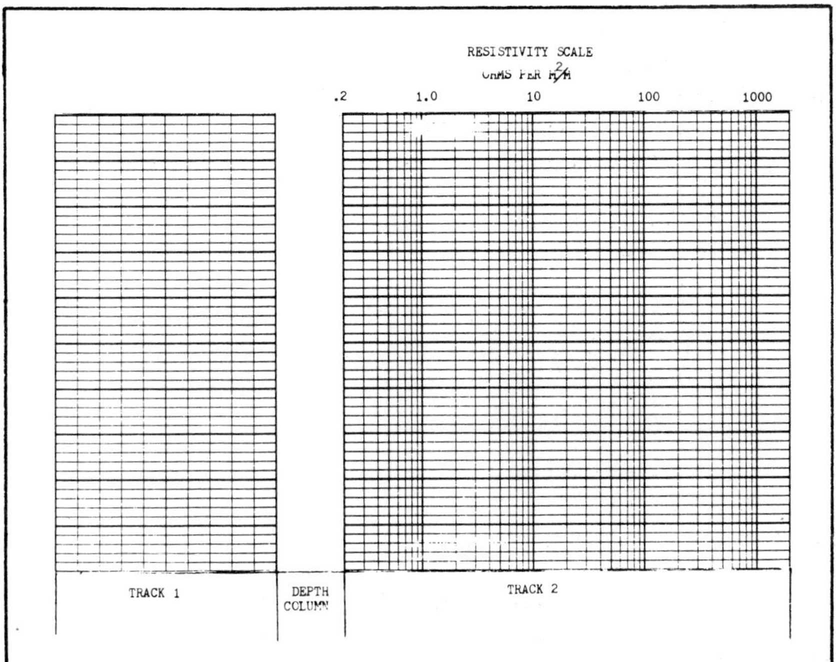

Logarithmic Well Log Grid
Figure 15

following list shows some of the more common logs of each type:

Conventional Logs:

- A. Flushed zone measurements.
  - 1- Microlog
  - 2- Minilog
  - 3- Contact Log
  - 4- MicroSurvey
- B. Deeper investigating logs
  - 1- Electrical Log
  - 2- Electrolog
  - 3- Electric Log

Focused Logs:

- A. Flushed zone measurements.
  - 1- Microlaterolog
  - 2- Proximity Log
  - 3- Minifocused Log
  - 4- FoRxo Log
- B. Deeper investigating logs.
  - 1- Induction - Electrical Log
  - 2- Dual Induction - Laterolog
  - 3- Dual Induction - Guard Log
  - 4- Laterolog
  - 5- Guard Log
  - 6- Dual Laterolog

Although there are many logs listed they are all designed to accomplish the same thing: to measure resistivity.

The presentation of resistivity measurements is standardized to the extent that they are always on the right side of the log. There are two standard grids, linear and logarithmic. Figure 14 shows the linear grid and Figure 15 is the logarithmic grid.

You will notice that both grids are divided by a blank vertical space to the left of center. This is called the depth column and is where the numbers are printed indicating the depths at which the measurements were made. On the linear grid the section left of the depth column is designated "Track 1" while those to the right are "Track 2" and "Track 3". On the logarithmic grid the left section is "Track 1" and the entire right section is "Track 2". Resistivity curves will always be presented in track 2 or tracks 2 and 3.

Scales denoting resistivity values will vary from area to area, but they will always have one thing in common: resistivity increases as the curves deflect toward the right. With the linear grid zero will be at the depth column for track 2, and at the left edge of track 3 for track 3. Full-scale deflection for each track may be 10 ohm-meters; 20 ohm-meters; 50 ohm-meters; 100 ohm-meters, or others. The value chosen will depend upon resistivity values expected. For example, Gulf Coast has low resistivity formations so the lower scales will be used in that area. Hard rock country like Mid-Continent or Rocky Mountains generally has high resistivity formations, so the higher scales will be used.

With the logarithmic grid the scale is pretty well standardized for all areas. At the depth column the scale starts at .2 ohm-meters (not zero) and progresses through cycles of 10 to 2000 ohm-meters at the right side of the grid. This allows a very broad range of resistivity values to be recorded without confusing extra scale traces that are necessary on linear grid logs when high resistivity values are encountered. Track 1 is reserved for other measurements, and will be discussed later.

The following figures are examples of the various logs presented in the previous list. Each curve is identified as to its depth of investigation.

Figure 16 is an example of a conventional flushed zone log such as Microlog, Minilog, Contact Log, or Microsurvey. This log is made with an instrument designed to be applied against the wall of the hole by a spring system. This allows the instrument to confine its measurements to within a few inches, therefore measuring just the flushed zone. Two simultaneous measurements are made, each with a different depth of investigation, but both are small. They are affected by the flushed zone and the mud cake, when present. Since the mud cake forms primarily on porous permeable

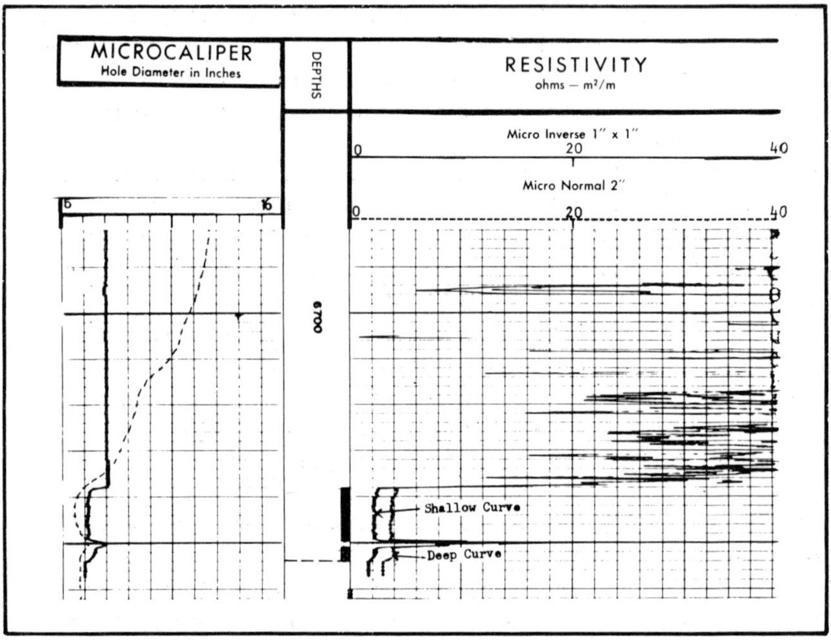

Conventional Flushed Zone Log
Figure 16

zones this log will identify such a zone. Such a zone is shown at depths 6738 to 6754′ by the separation between the curves. Separation such as this is called "positive separation" and is the condition generally created by the presence of mud cake. Unusual circumstances can cause variations from this, but they are not the norm. This log will be discussed more in detail in the chapter on porosity measurements.

Figure 17 is a conventional Electrical Log. The shallow investigation curve measures predominately in the invaded zone. The deep investigation curve measures much further into the formation and attempts to determine the true resistivity of the non-invaded zone. The medium investigation curve will measure somewhere in between the other two, and its measurement will depend on the depth of invasion. If invasion is very deep, then this curve will more nearly resemble the shallow investigation curve. If invasion is shallow then this curve will favor the deep investigation curve.

This would seem to be an ideal arrangement, and might be so if it were not for problems created by variations in hole size, mud resistivity, formation resistivity, formation thickness, invasion, and other conditions that can cause the values measured by these curves to be altered from their true values. Techniques to correct these values to overcome the adverse effects were developed and used for quite a time. A much better solution came along with the development of focused logs.

Figure 18 shows a focused log designed to measure within the flushed zone, like the Microlaterolog. It is also made with an instrument that is applied to the wall of the hole with a spring system. This log consists of a single measurement so that the flushed zone resistivity is read directly off the log, thus simplifying the technique and increasing the accuracy. It is recorded on a hybrid scale in track 2, which is linear for the first five divisions, then compressed to infinity in the next five divisions. The curve in track 3 is an amplified linear scale which permits accurate determination of resistivity values in the low range. This log also will be discussed in detail in the chapter on porosity.

Figure 19 is an Induction - Electrical Log, which retains the shallow

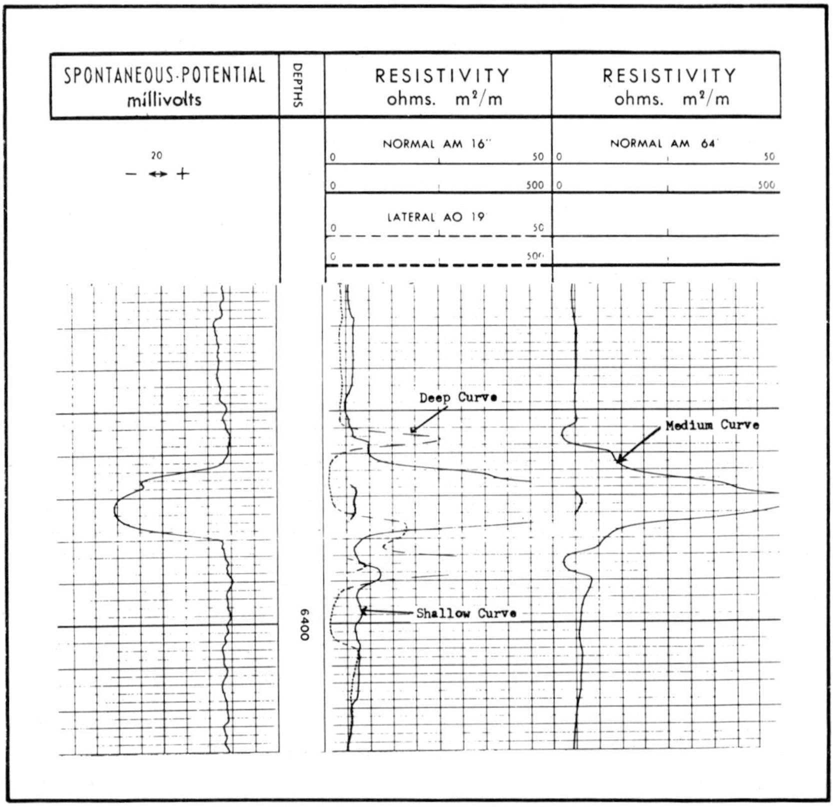

Conventional Electrical Log
Figure 17

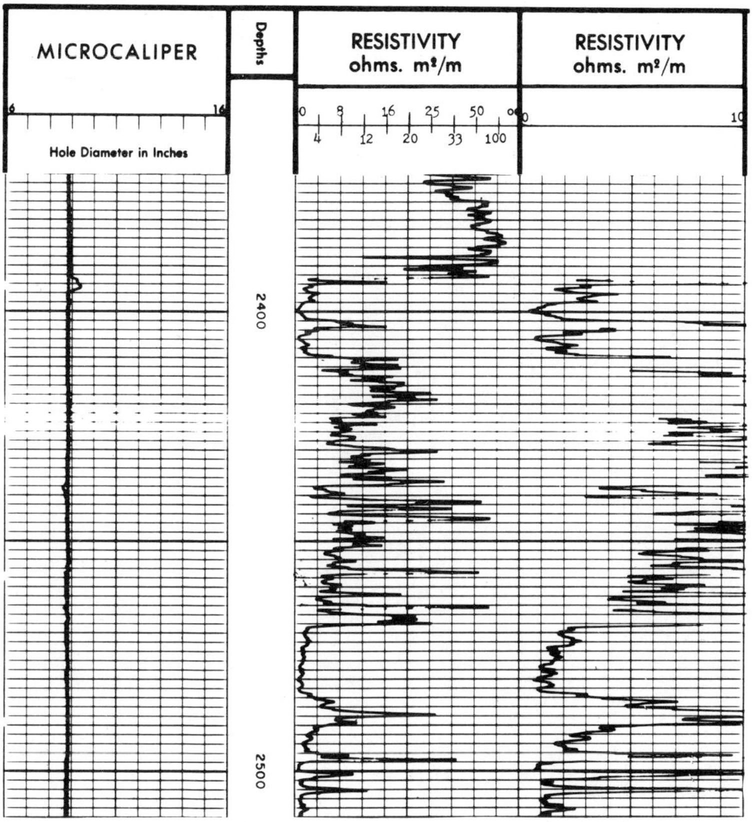

Focused Flushed Zone Log
Figure 18

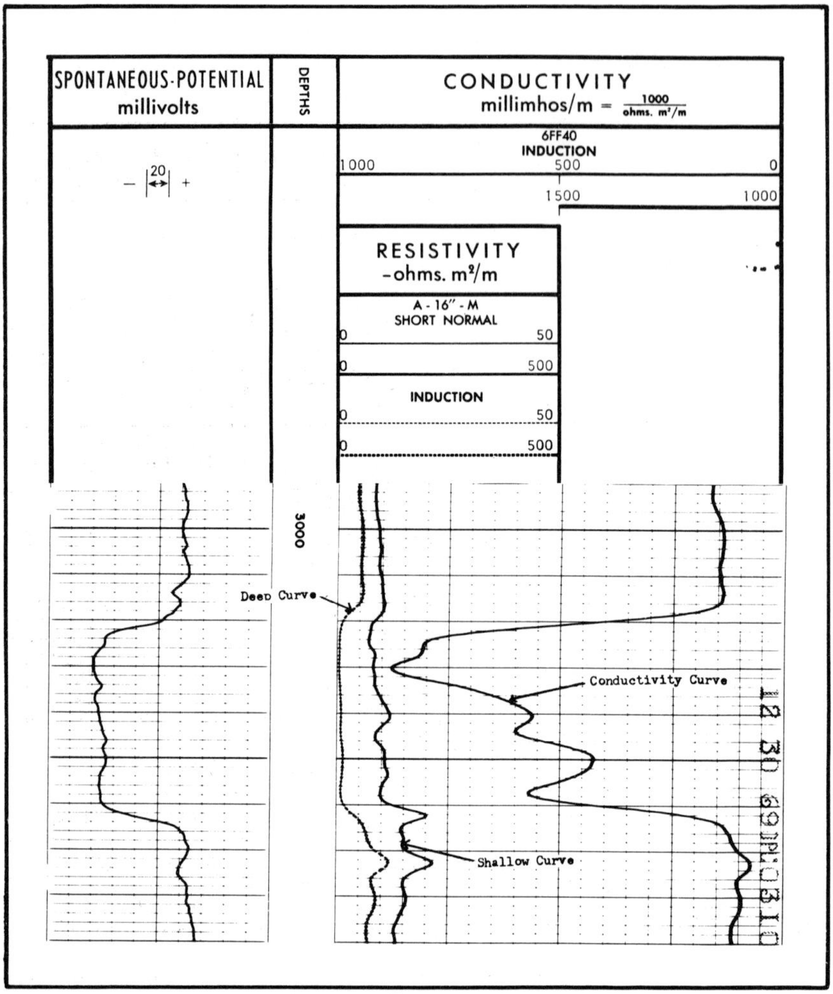

Induction—Electrical Log
Figure 19

investigating curve from the conventional Electrical Log and combines it with an induction type focused deep investigation curve. There is no medium investigation measurement on this log. A note of interest is that the induction log actually measures formation conductivity rather than resistivity. Since they are the reciprocal of each other the same information is obtained. The conductivity curve is presented in tracks 2 and 3. The scales for this log are somewhat different than those shown in the previous examples. The reason for the difference is that the unit of conductivity is a mho (ohm spelled backwards). With the range of conductivity generally encountered fractions of mhos would be necessary to indicate the values measured. In order to avoid fractions milli-mhos are used in the scaling (thousandths of mhos). Zero conductivity is at the right edge of the log and increases to 1000 milli-mhos at the depth column. Consequently, full-scale deflection ranges from 1 ohm-meter at the depth column to infinity at the right edge of the grid. Keep in mind that resistivity equals 1000 divided by the conductivity in milli-mhos. In order to avoid having to make this calculation each time a resistivity value is wanted, the control equipment in the logging truck does this simultaneously and continuously as the log is run. The resistivity values are presented in track 2 as a dashed curve. It is scaled as usual with zero at the depth column and increases with deflection to the right. There will be more discussion in detail in the section on analysis of logs.

The log in Figure 20 is a multi-curve log with all curves of the focused type. This is the Dual Induction - Laterolog or Dual Induction - Guard Log. Each of these logs is presented on the logarithmic grid. The shallow investigating curve is the Laterolog or Guard Log, and the medium and deep investigating curves are induction type. The surface control equipment converts all the induction measurements into resistivity so that all curves are presented on the grid as OHMS.M^2/M. There will be more detail on this log in the analysis chapter.

Figure 21 shows a Laterolog or Guard Log. This is a single curve log even though there are two curves presented. Please note the scaling. There

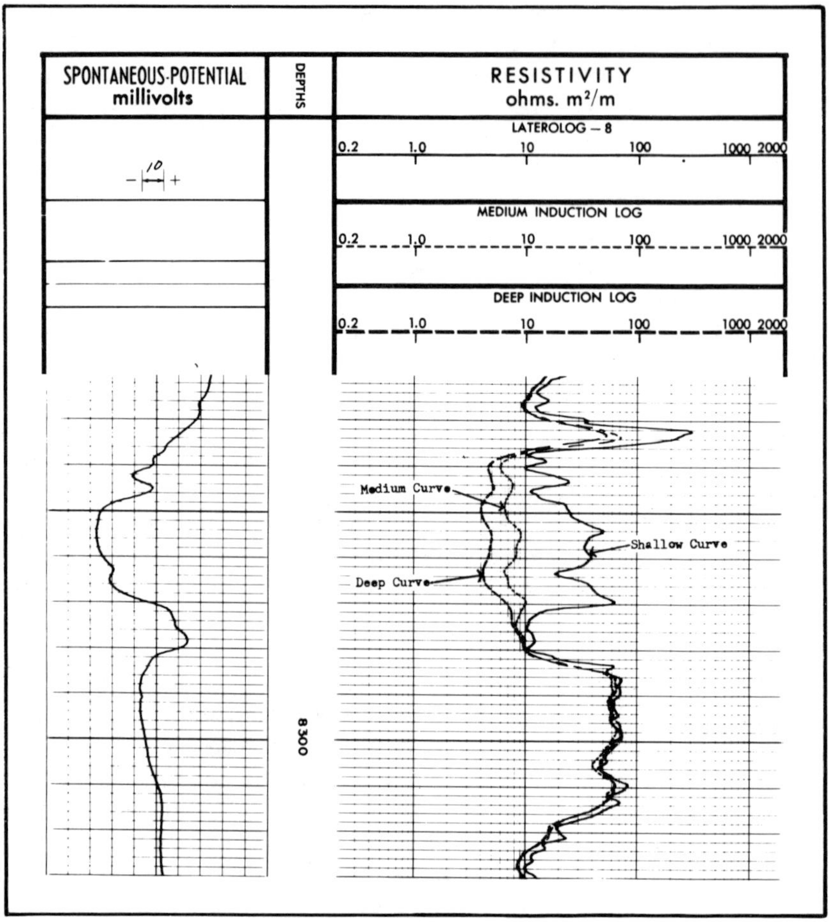

Focused Multi-Curve Log Dual Induction—Laterolog
Figure 20

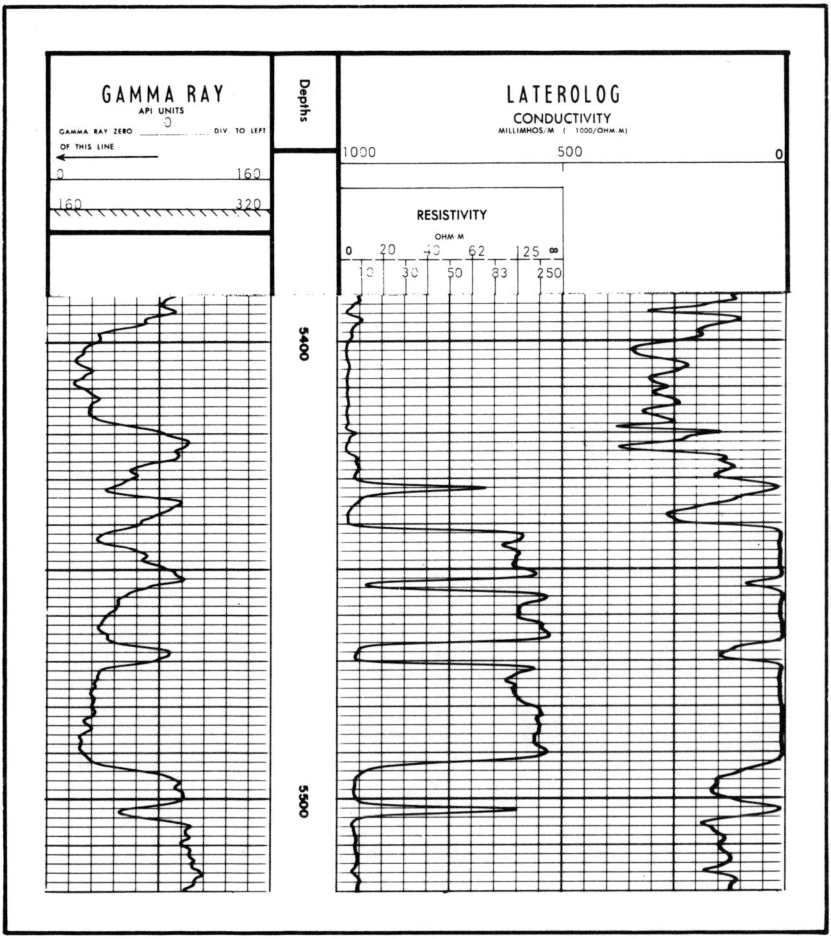

Focused Deep Investigating Log—Laterolog or Guard Log
Figure 21

is a curve in track 2 which is resistivity recorded on a hybrid scale, linear for the first five grid divisions then compressed to infinity in the next five divisions. The curve to the right is the same log presented in milli-mhos of conductivity with zero at the right edge of track 3. This permits an accurate determination of resistivity regardless of the value. This is especially valuable, since the Laterolog or Guard Log is quite often run where resistivities are very high. Since there is only one curve it is difficult to determine the effects of invasion. As an aid to overcoming this problem the Dual Laterolog was developed.

The Dual Laterolog is shown in Figure 22. Sometimes the Dual Laterolog will be recorded with a focused flushed zone log as shown in Figure 22 - A. The purpose of this is to give three depths of investigation which permits a better solution of the invasion problem.

§ 1.03 Selection of Resistivity Logs

Since all well logs must be run in the bore hole it is apparent that the bore hole environment will influence the selection of logs used to measure resistivity.

As mentioned previously almost all wells are drilled with the rotary method, and most of these use some sort of fluid as the circulating medium. Some will use air or gas. The most common rotary fluid is mud made up of water mixed with clays and chemicals. Some will use oil or oil mixed with clays and chemicals. In the water-based mud systems the make up water may be either fresh or salty. The fresh water mud will be considerably more resistive than salt water mud. In logging terminology mud resistivity is compared to formation water resistivity to determine whether it is considered fresh or salty. Mud is considered fresh when its make up water is at least five times more resistive than the formation waters, a greater ratio is preferred. Mud is considered to be salty when its make up water approaches the same resistivity as formation waters.

Conventional resistivity measurements are made with instruments that require an electrical connection between the bore hole instrument and the

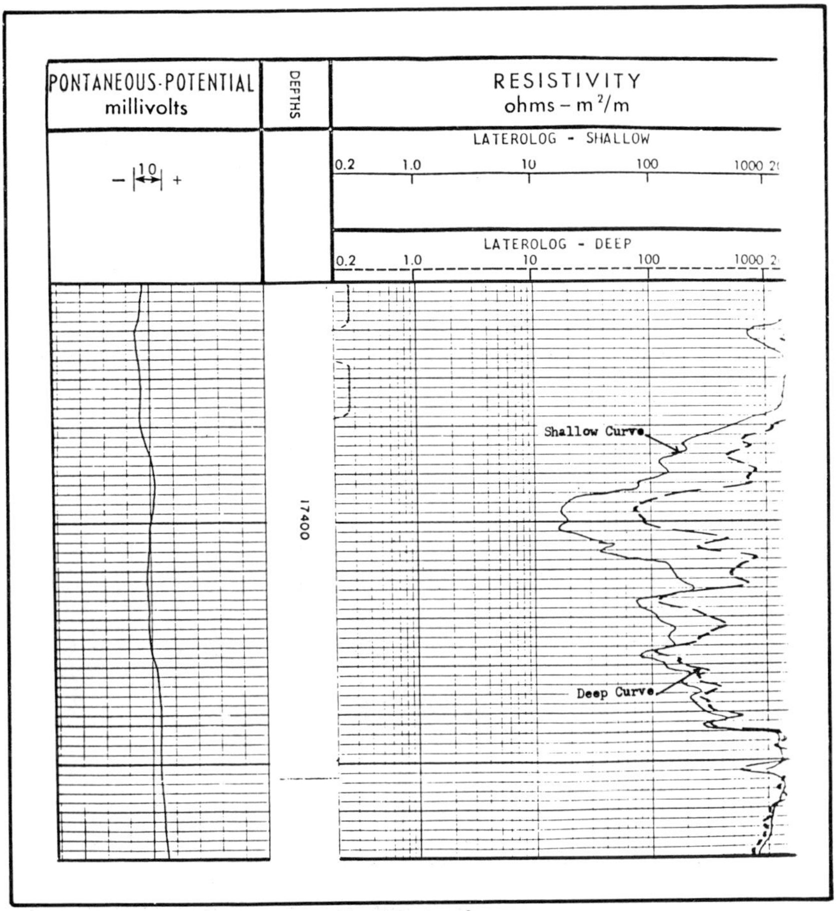

Dual Laterolog
Figure 22

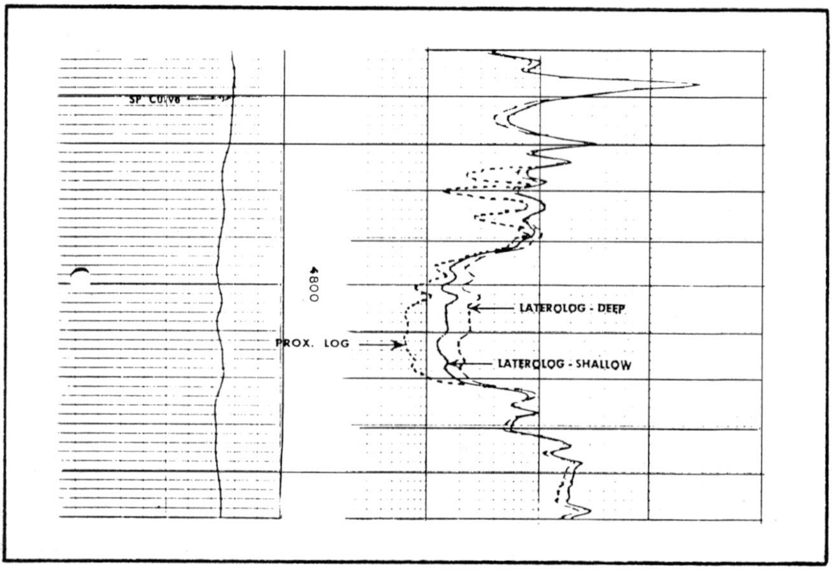

Dual Laterolog With Focused Flushed Zone Log Added
Figure 22A

formations. This is accomplished with a water-based mud system. In order for these instruments to operate at their maximum efficiency the mud must be fresh. Fresh mud confines the logging current flow to the vicinity of the electrode systems involved. If the same instruments are used in salty mud the measurements obtained will be badly distorted or altered from their proper values. Electricity will flow through the path of least resistence, so in salty mud the logging currents will flow up and down the hole rather than into the formations as they should. This causes the distortion of values. In order to overcome this problem focused logs were developed. The first of these was the Laterolog and Guard Log. They use an electrode arrangement that prevents the flow of logging currents up and down the hole and forces them to flow into the formations. Consequently, more accurate measurements are made under these circumstances.

Conventional resistivity measurements also are affected by formation thickness, and formation resistivity (especially high resistivity formations). Focused logs helped to overcome these problems also. Induction type focused logs use electromagnetic waves in order to make their measurements. This permits their use in non-conductive bore hole environments such as air or gas, or in oil-based mud systems as well as water-based mud systems. Because of all these advantages nearly all resistivity logs now are made with focused logging instruments.

Bore hole environments and logs which may be used are as follows:

§ 1.03(a) Water-Based Mud Systems

Fresh mud

Electrical Log - shallow, medium, and deep curves - each of which may require corrections to achieve proper resistivity of zone investigated.

Induction - Electrical Log - shallow and deep curves - difficult to detect invasion effects - shallow curve may need correction.

Dual Induction - Laterolog - (Dual Induction-Guard) - shallow, medium, and deep curves - Laterolog and Guard curves are especially designed to investigate the invaded zone. Since all the curves are focused, corrections are minimal. Provides most accurate resistivity measurements.

Salty mud

Laterolog (Guard Log) - one curve - deep investigation. Difficult to detect and correct for invasion effects.

Dual Laterolog - medium and deep curves - helps detect invasion. May include very shallow curve for more accurate analysis.

§ 1.03(b) Oil or Oil-Based Mud Systems

Induction type logs - generally one deep curve. Practically no invasion with this hole fluid to affect determination of true resistivity.

§ 1.03(c) Empty Holes (Air or Gas)

Induction type logs - generally one deep curve. No fluid to invade formation and affect resistivity values.

Along with bore hole environment the selection of logs may depend on other things, among which are: depth of hole, number of possible pay zones, thickness of possible pay zones, lithology of pay zones, structural position (oil/water contacts), quality of the mud program, whether the well is an in-fill well, development well or wildcat well, and economics.

Each of these factors will be discussed at length in the chapter where complete logging programs are developed.

Let us now turn our attention to the next electrical characteristic of rocks, spontaneous potential.

§ 1.04 Spontaneous Potential

The spontaneous potential (SP) curve is a measurement of potential (voltage) variations in the bore hole, recorded continuously vs depth, caused primarily by the interaction of the mud filtrate in the hole and the formation waters. There are other factors that help cause the SP, but the ratio of mud filtrate resistivity (R_{mf}) to the formation water resistivity (R_w)

is the most important.

In Figure 23 we have an insulated container divided into two sections by a thin membrane, whose only function is to separate two solutions of different salinities yet allow ionic transfer between the two solutions. Both the solutions contain only sodium chloride (N_aC_1), however one has a very high concentration and the other a very low concentration. Sodium (N_a)

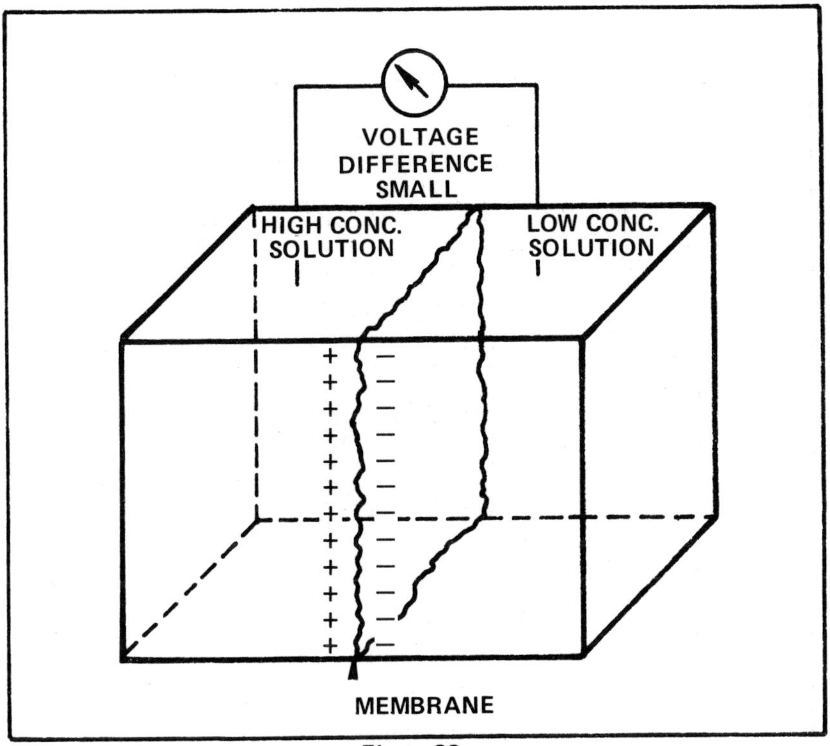

Figure 23

ions are positive ions, and clorine (C_1) ions are negative ions. Since the two solutions are in direct contact (through the membrane in Figure 23) the ions are free to transfer from one solution to the other. The negative ions have greater mobility so they will move from the highly concentrated solution to the lower concentrated solution. This causes a high concentration of negative ions on one side of the membrane and high concentration of positive ions on the other side. This creates a battery effect between the two

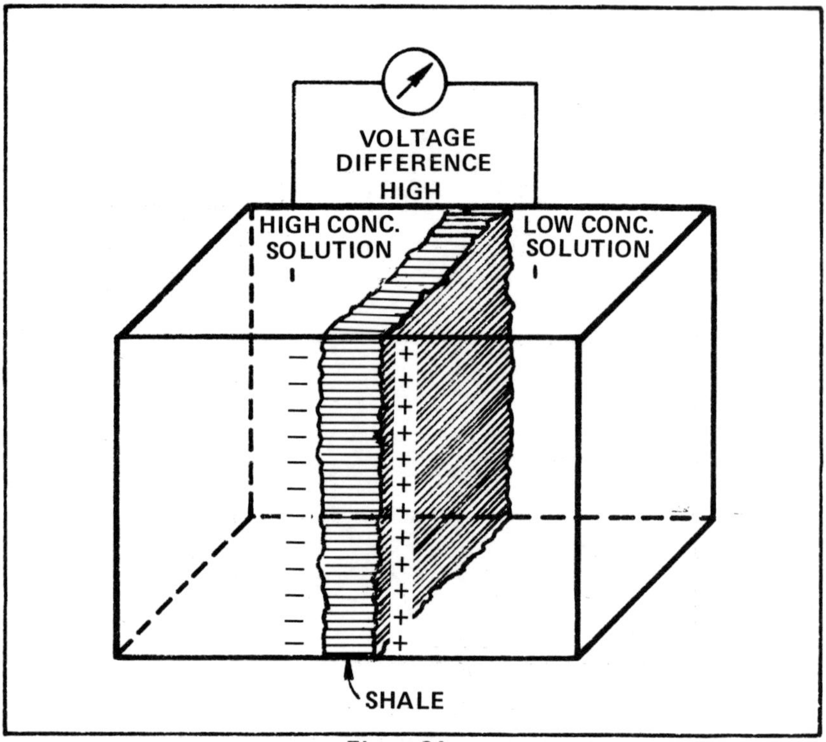

Figure 24

solutions. The potential (voltage) difference across the membrane will be small, only a few millivolts.

Figure 24 shows the same container and two solutions. The membrane has been replaced by a shale partition. Shale is permeable to positive ions (Na^+) but impervious to negative ions (Cl^-). There is now a concentration of positive ions on one side of the shale and negative ions on the other. Notice that the concentrations are now opposite of what they were when the membrane separated the solutions. Again, a battery effect is created but with the polarity reversed, and the difference of potential is much greater, many millivolts.

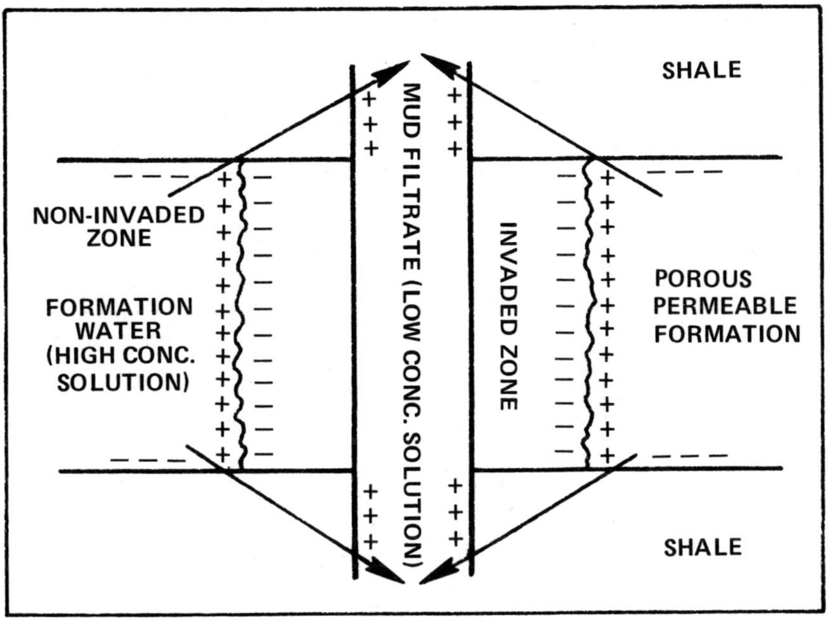

Figure 25

Now let's see how we can have both these conditions simultaneously in a bore hole. Figure 25 shows a porous permeable formation with shales above and below. The two solutions are the mud filtrate (low NaC_1 solution) and the formation water (high N_aC_1 solution). At the boundary of the invaded zone the two solutions are in direct contact and the ions are free to transfer. There will be concentrations of positive and negative ions as shown. The two solutions are in contact through the shale at the top and bottom boundaries of the formation as shown by the arrows. The shale passes the N_a+ions but not the C_1-ions. The concentration at the porous bed/shale contact will be negative. The concentration at the shale/mud

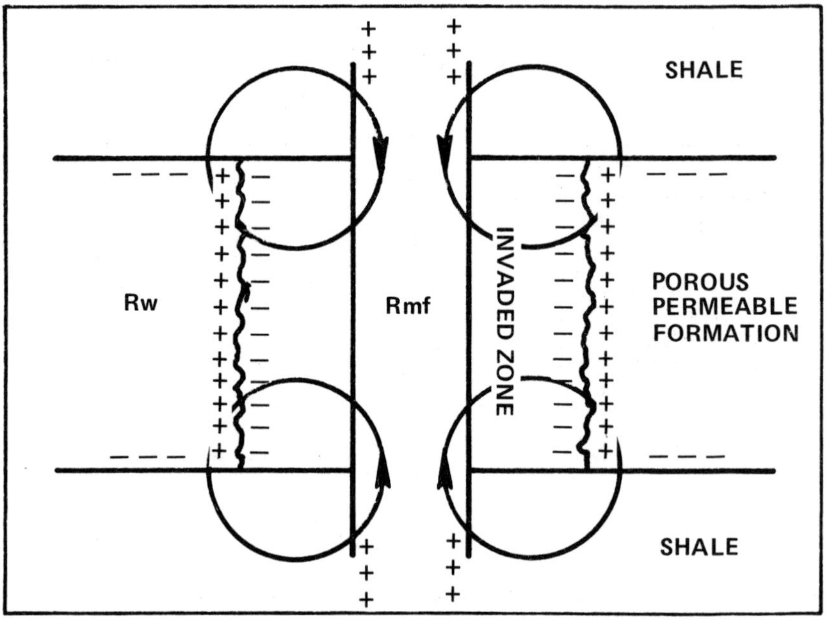

Figure 26

column contact will be positive. As a result, we have two battery effects and the polarities are reversed.

Figure 26 shows the resulting current flow around the junction of the porous formation, the shale, and the mud column. Current flow around this junction must travel through the mud column, upward at the bottom junction, and downward at the upper junction. There will be a different

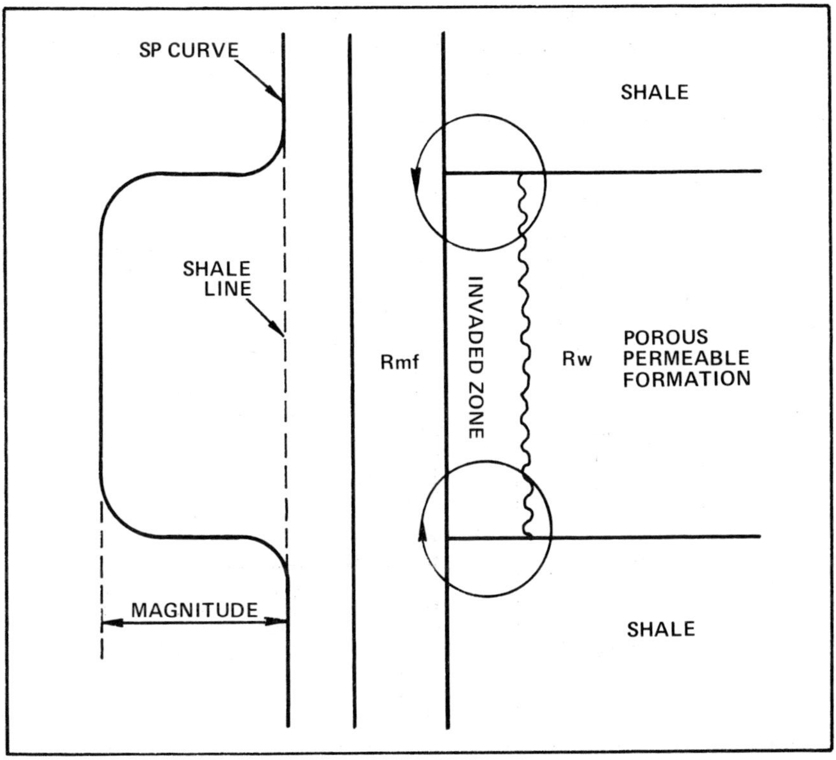

Figure 27

potential at each foot of depth in the mud column and the polarities will be opposite at the two boundaries. An electrode moved up (or down) through the mud column will measure this difference of potential. The resulting measurement is the SP curve. It will appear as shown in Figure 27.

The SP curve is recorded in track 1 of the log (left of the depth column). Deflections are measured from the shale line, which is the value present in front of the shales in the well. Most of the time this is the value nearest the depth column and is adjusted by the logging engineer. Deflections to the left from the shale line are "negative SP" and those to the right from the shale

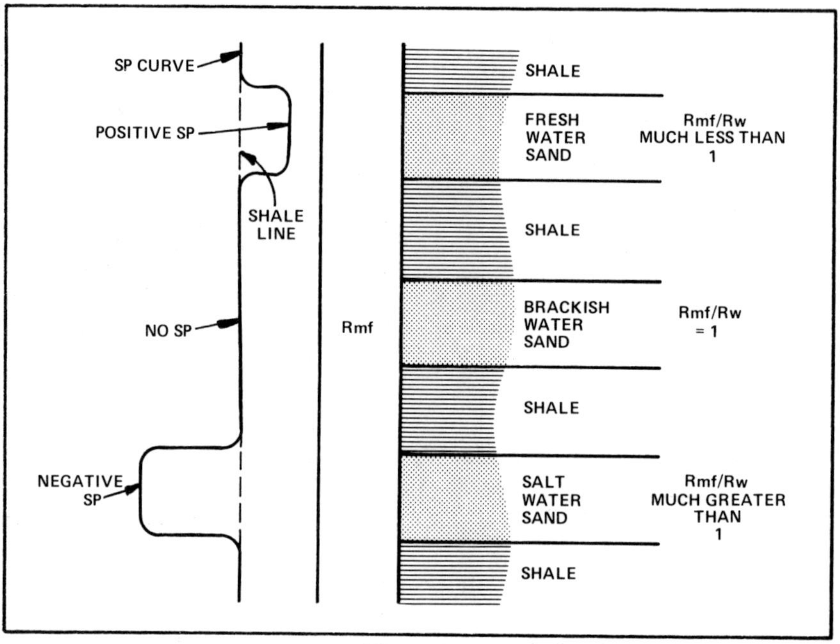

Figure 28

line are "positive SP". Typical fresh mud SP is shown in Figures 17-19 and 20. Typical salty mud SP is shown in Figures 22 and 22-a.

The ratio of the mud filtrate resistivity (R_{mf}) to the formation water resistivity (R_w) controls the polarity and magnitude of the SP curve. If the ratio is large the SP will be negative have a high magnitude of deflection. This is the case where the mud filtrate is fresh and the formation water is salty. If the ratio is near to unity (1) there will be no SP. This is the case where both the mud filtrate and formation water are salty (they could both be fresh, also). If the ratio is a fraction the SP will be positive. This is the case where the mud filtrate is saltier than the formation water, generally found in front of fresh water sands. These conditions are shown in figure 28.

The SP responds to the presence of a porous permeable formation, but the magnitude of the SP deflection has no relationship to the amount of porosity or permeability. The porosity is where the formation water is, and there need only be enough permeability to allow ionic transfer.

The mathematical relationship for the SP is:

$$\text{SPONTANEOUS POTENTIAL} = -K \log \frac{Rmf}{Rw} \qquad \text{(SP-1)}$$

K is a coefficient which is proportional to temperature, and for Na C1 solutions equals 71 @ 77° F. For solutions containing substantial amounts of other salts K may vary.

The static SP (SSP) is the maximum SP measured opposite a thick water sand for a given Rmf/Rw ratio.

The SP curve is useful in many ways. It will:

1- Detect porous permeable formations
2- Locate their boundaries
3- Permit a determination of Rw when no other information is available
4- Indicate shaliness of a formation
5- Assist in correlation

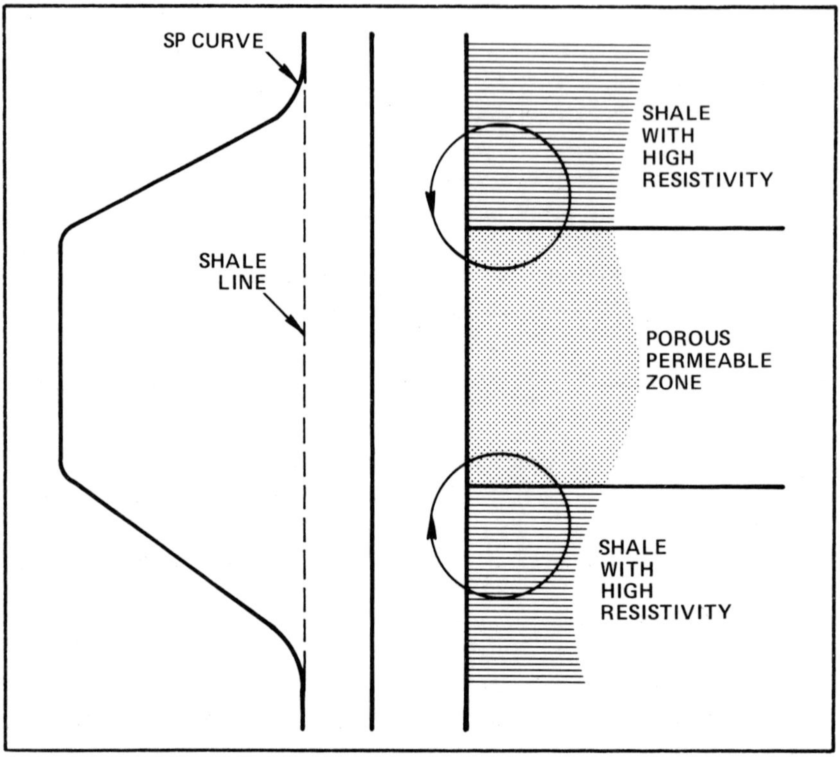

Porous Zone Between Resistive Shales (Rmf/Rw Large)

Figure 29

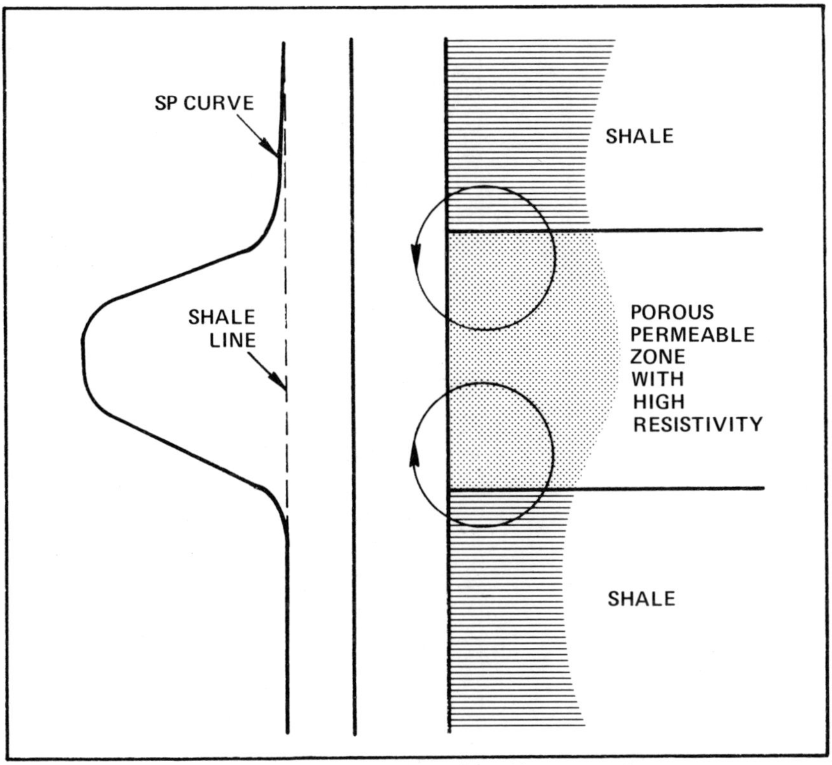

Resistive Porous Zone (Rmf/Rw Large)

Figure 30

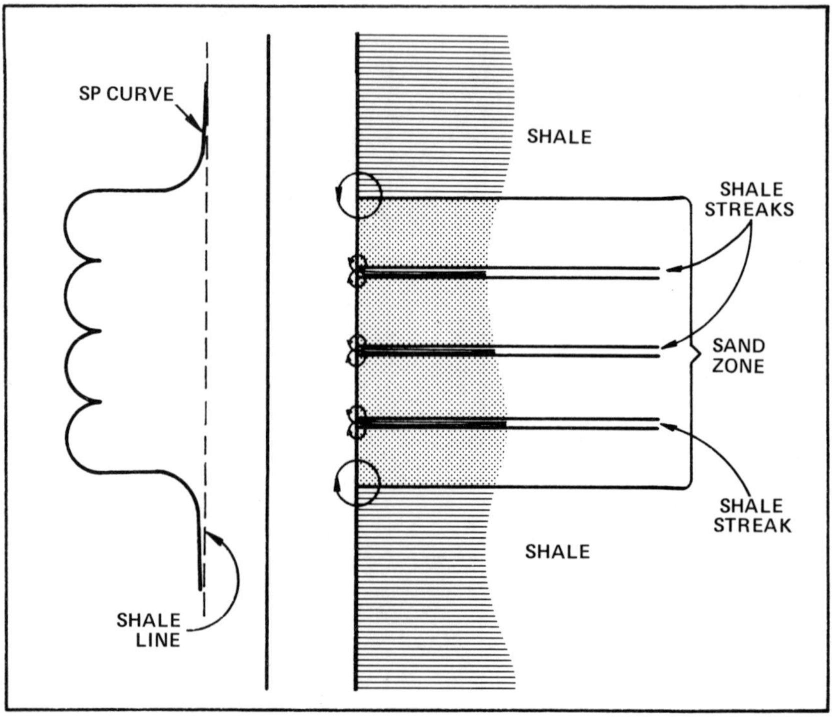

Sand With Shale Streaks (Rmf/Rw Large)

Figure 31

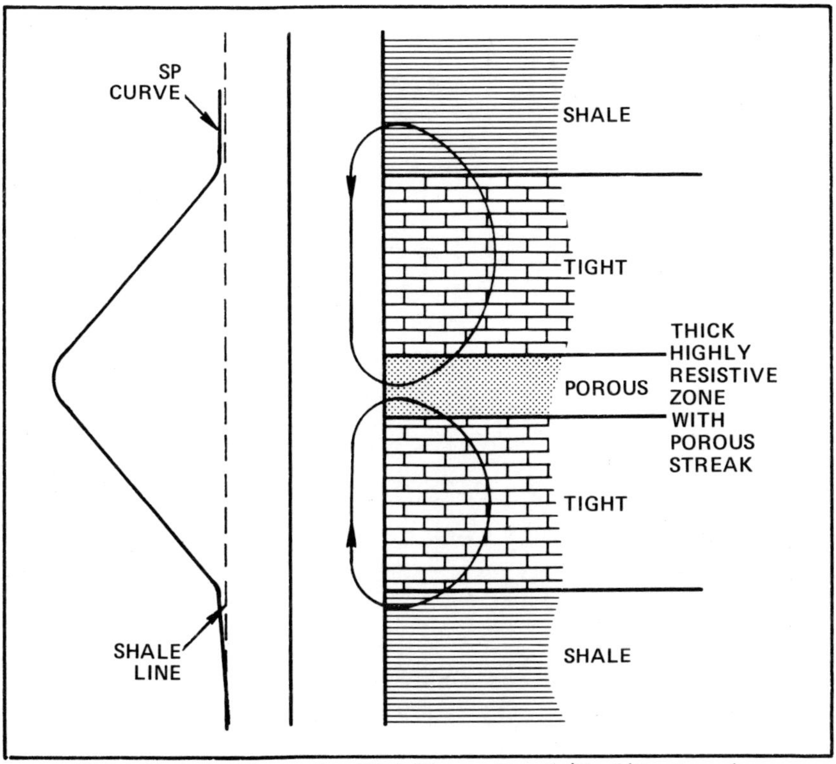

Thick Resistive Formation With Porous Streak (Rmf/Rw Large)

Figure 32

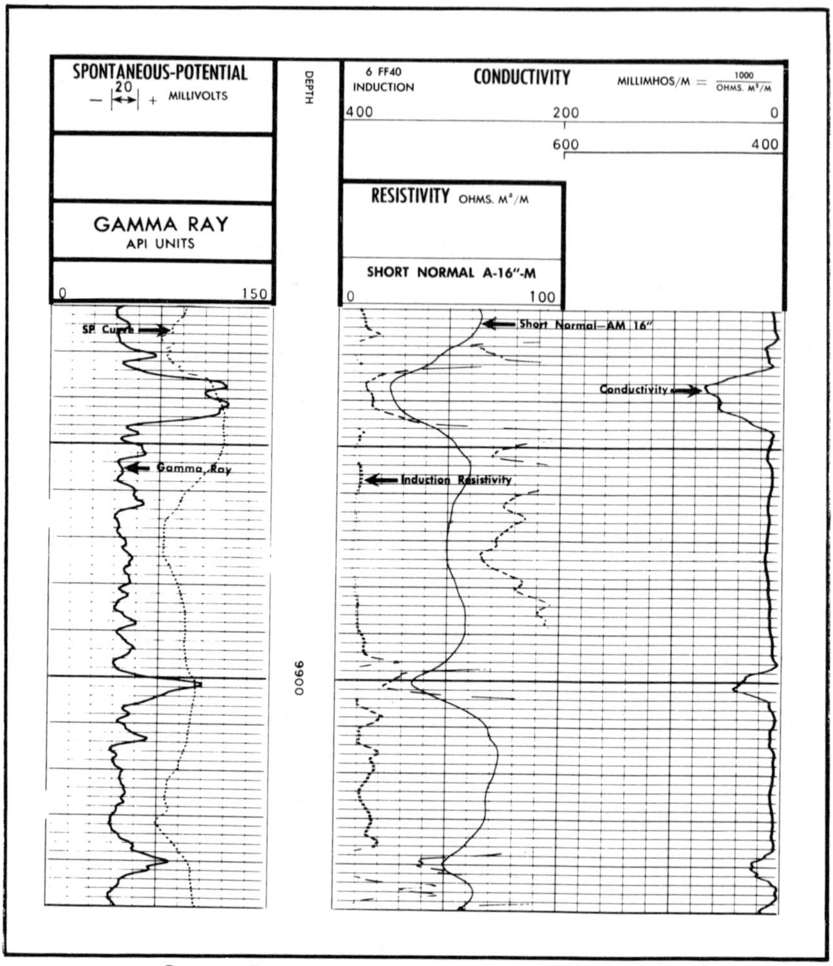

Gamma Ray Curve Recorded With An SP Curve
Figure 33

The shape of the SP curve often can be very useful. Some of these conditions are shown in Figures 29, 30, 31 and 32. In each case the path of current flow is shown along with the resulting curve shape.

There are some factors that may affect the magnitude of the SP curve other than the Rmf/Rw ratio. Among them are enlarged hole, deep invasion, formation thickness (thin beds), formation resistivity (high), and shaliness. The net effect on the SP is to reduce the magnitude. Very often the magnitude of the SP in shaley sands is compared to the magnitude of the SP in clean sands to determine the amount of shale in the shaley sand. This is helpful in some interpretation techniques.

§ 1.05 Radioactivity

At the time of deposition various rocks were contaminated with radioactive material such as potassium, thorium, and uranium, among others. As a general rule shales contain much radioactive material, while reservoir rocks such as sandstones, limestones, and dolomites contain small amounts.

In nature some radioactive elements emit spontaneous bursts of high energy electromagnetic waves called gamma rays. A device that can detect and measure the intensity of gamma rays and record them continuously versus depth would produce a gamma ray log.

Therefore, a Gamma Ray Log is a measurement of the natural radioactivity of the formations traversed by the bore hole. The formations with high radioactivity are shales, and those with low radioactivity are, or may be, potential reservoir zones.

The Gamma Ray Log is nearly always recorded in track 1 of the log. It is scaled so that low radioactivity is near the left side of the track and radioactivity increases toward the depth column. Consequently, its anomalies are similar to those of the SP curve. In fact, the Gamma Ray Log frequently is used to complement the SP or as a substitute for the SP where the SP is not definitive. Figure 33 shows a typical gamma ray curve recorded with the SP curve. (A Gamma Ray Log may also be seen in track 1 of Figure 21.)

The Gamma Ray Log may be made in almost any bore hole environment, such as empty hole, oil or oil based mud filled hole, water based mud filled hole (either fresh or salty), or cased hole. It can be run in conjunction with many other logs to facilitate their analyses.

Among the many uses of the Gamma Ray Log are:

1- Definition of lithology changes, especially shales from reservoir rocks
2- Indication of shaley content of reservoir rocks
3- Correlation for mapping and depth control (especially completion services)
4- Detection and evaluation of radioactive minerals
5- Detection of radioactive markers used in completion work
6- Detection of tracer material used in completion work.

Chapter Two

POROSITY

Porosity, the amount of void space in percent of total volume of rock, is the storage space in which hydrocarbons can accumulate. The knowledge of an accurate value for porosity is important for a proper evaluation of a reservoir.

Since the very early days of logging, a method to measure porosity in place has been the subject of continuing efforts. One of the first such efforts was the Neutron Log, which at its best was qualitative. Research continued until several porosity measuring systems were developed.

§ 2.01 Porosity Measuring Logs

These porosity measuring logs are as follows:

1- The MicroLog (Contact Log, MiniLog, etc.)
2- The MicroLaterolog (FoRxo Log, Micro Focused Log, etc.)
3- The Sonic Log (Acoustic Log, etc.)
4- The Bore Hole Compensated Sonic Log
5- The Density Log (Densilog, etc.)
6- The Compensated Density Log
7- The Sidewall Neutron Log
8- The Compensated Neutron Log

As discussed previously in the resistivity section, the micro devices (Numbers 1 and 2) are designed to measure within the flushed zone of a porous permeable formation (the few inches away from the bore hole wall where mud filtrate has flushed the original formation fluids). In order to eliminate bore hole effects from this measurement the instruments are designed so that the insulated pad that contains the electrode system is pressed on the wall of the hole as shown in Figure 34. (Examples of these logs are shown in Figures 16 and 18.)

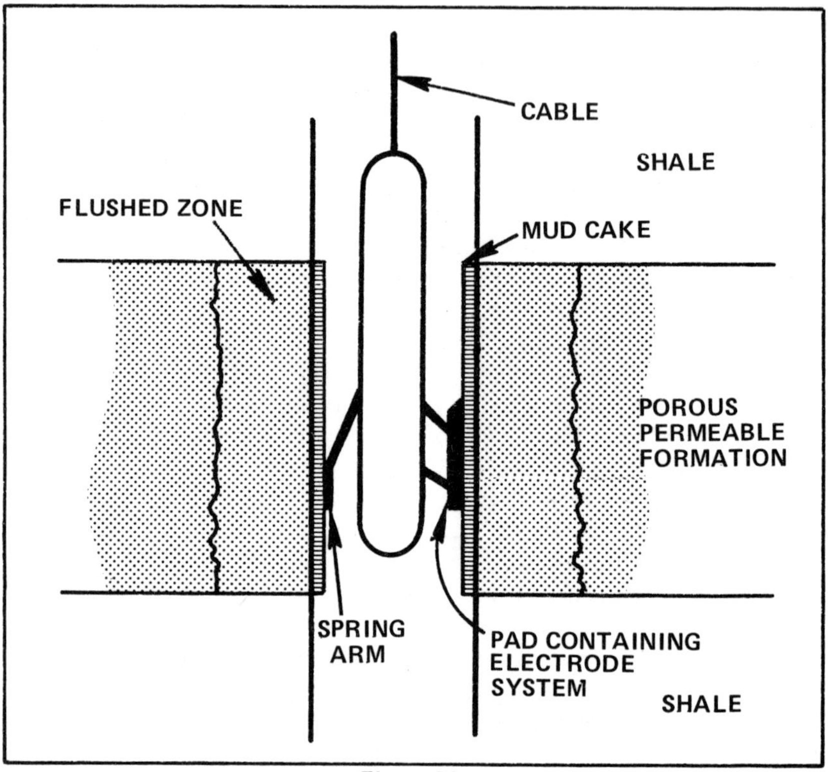

Figure 34

The resistivity of the flushed zone (Rxo) is influenced by the formation resistivity factor (F), the mud filtrate resistivity (Rmf), and whatever original formation fluids are left. In the case of a hydrocarbon-bearing reservoir there will be some of the oil or gas left in the flushed zone. This is called residual hydrocarbon saturation (Shr). Since the resistivity must be measured with an instrument in the bore hole with a pad pressed against the wall of the hole, the mud cake thickness (h_{mc}) and resistivity also must be considered. Therefore we have five factors that will have an effect on the resistivity measurement. To recap, these are: F, Rmf, Rmc, h_{mc}, and Shr. Of these, only two may be known with any degree of accuracy. They are Rmf and Rmc, which can be measured at the surface using a filter press and a sample of the drilling mud. There is no assurance that these measured values will be the same as those encountered down in the well. The mud cake thickness (h_{mc}) can be determined with reasonable accuracy from the caliper log that is run simultaneously with the resistivity measurements. Residual hydrocarbon saturation (Shr) must be estimated. Formation factor (F) is what we want to determine, because from F we can estimate porosity using a relationship pertaining to the lithology.

Consider now that when a log has been recorded we have Rxo values for each porous permeable formation included in the logged interval. Using a mathematical procedure and the best values obtainable for the variables F is determined for each porous permeable formation of interest. From these F values porosity is derived. This technique will be demonstrated in the section on Analysis Techniques.

Derivation of porosity from Rxo values leaves a lot to be desired so far as accuracy is concerned. Because of this research continued searching for a method of measuring porosity which would give better results.

§ 2.01(a) Acoustic Logs

The next step was to take advantage of geophysical knowledge of the acoustical characteristics of rocks. The speed with which rocks will transmit sound depends on their elastic properties. More elastic rocks such as carbonates will transmit sound much faster than less elastic rocks like shales. Sandstones are less elastic than carbonates, but much more so than shales. The elasticity will vary as the porosity varies.

The Acoustic or Sonic Log is a continuous measurement versus depth of the amount of time it takes for sound to travel through one foot of rock. A device that will make these measurements is shown in Figure 35.

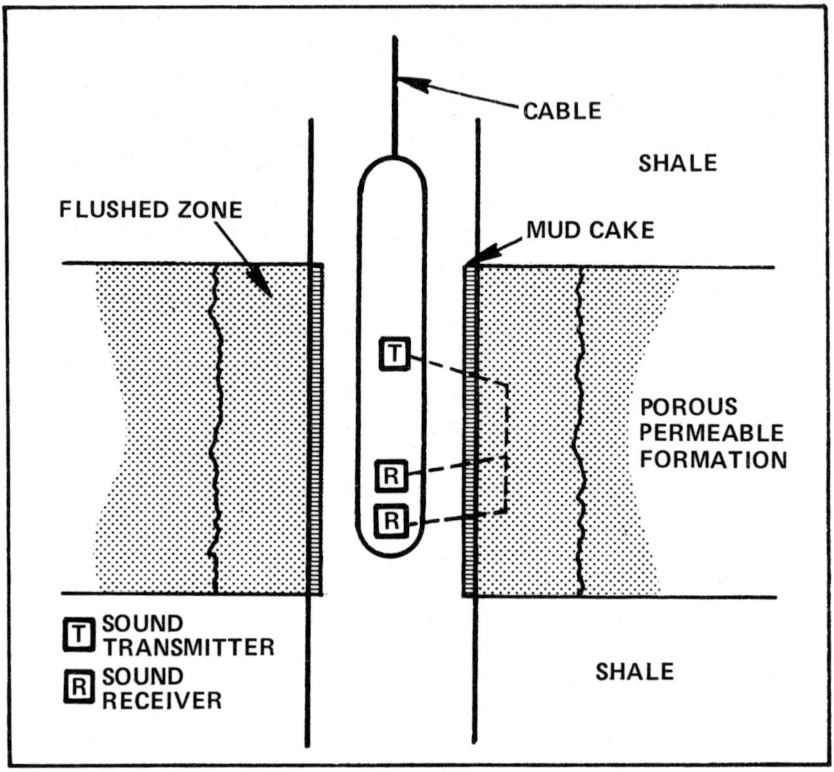

Figure 35

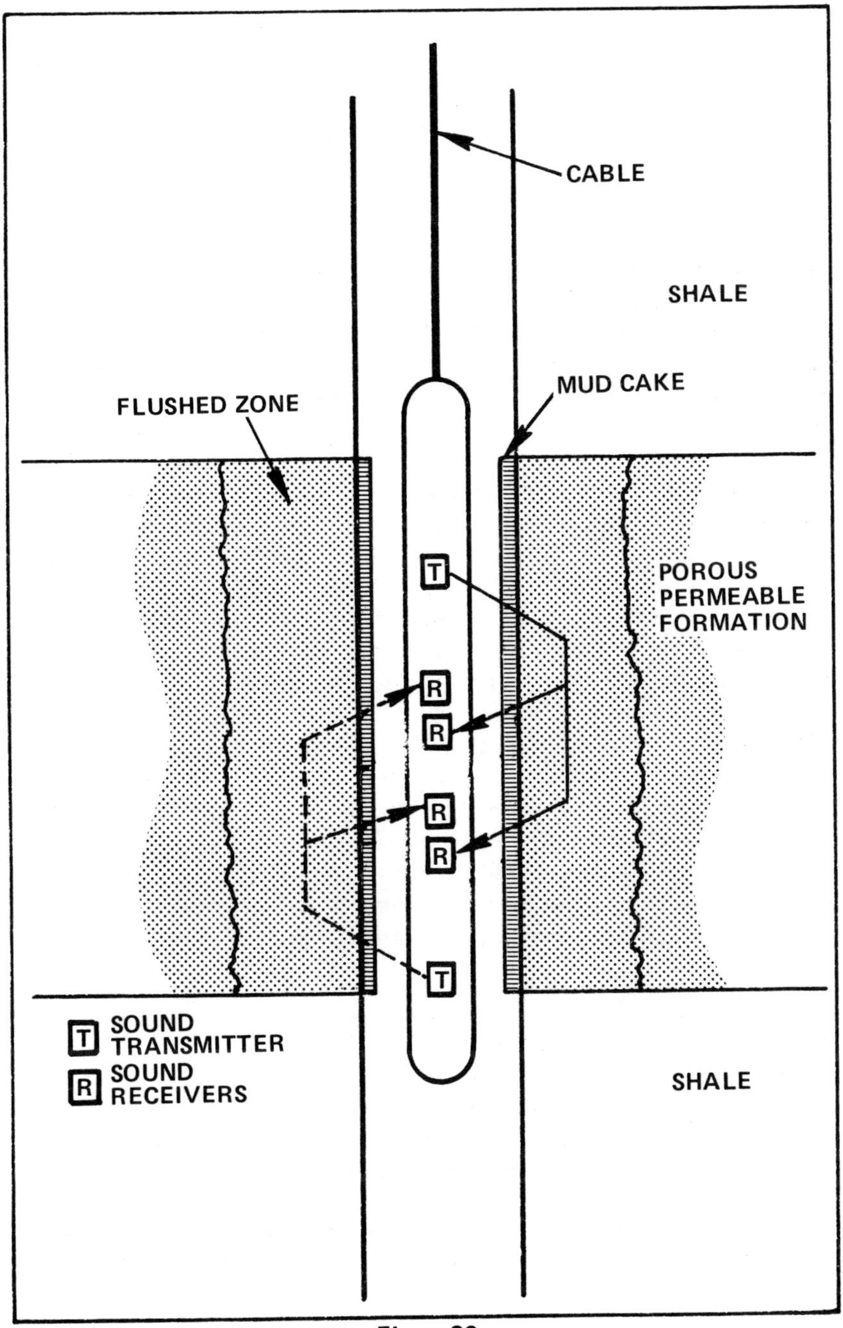
CABLE
SHALE
MUD CAKE
FLUSHED ZONE
T
POROUS
PERMEABLE
FORMATION
R
R
R
R
T
T SOUND
TRANSMITTER
R SOUND
RECEIVERS
SHALE

Figure 36

A ticking sound of constant frequency and strength is emitted from the transmitter. The sound will travel through the mud and then up and down through the rock near the bore hole. The sound that travels down toward the receivers will be "heard" by the first receiver and then by the second receiver. The instrument is designed to measure the time it takes for each receiver to "hear" the sound. The length of time taken to the first receiver is subtracted from the length of time taken to the second receiver. The difference is the time it takes for the sound to travel between the receivers and is related to velocity per foot of rock. The instrument shown in Figure 35 could produce time measurements that are affected by bore hole size variations. To overcome this problem the Bore Hole Compensated Sonic Log (BHC) was developed. This system uses two sets of transmitters and receivers as shown in Figure 36. Time measurements alternate between each set. The resulting time interval recorded on the log will be the most accurate value obtainable.

Since sound travels through rock very fast the time is measured in microseconds (millionths of seconds). Consequently the Sonic Log is scaled in Interval Transit Time in Microseconds per foot of rock. A typical Sonic or Acoustic Log is shown in Figure 37.

Mr. M. R. J. Wyllie with Gulf Research and Development Corporation proposed the time average relationship that has proven to be very successful in solving the Sonic Log for porosity. It is:

$$\emptyset = \frac{\Delta t_{log} - \Delta t_{matrix}}{\Delta t_{fluid} - \Delta t_{matrix}}$$

Where: Δt_{log} = travel time measured by the log

Δt_{matrix} = travel time through the rock at "O" porosity

Δt_{fluid} = travel time through the fluid in the porous space generally water (mud filtrate).

The travel time through rock at "0" porosity will vary, depending on the type of rock involved. Some common rock types and their velocities and

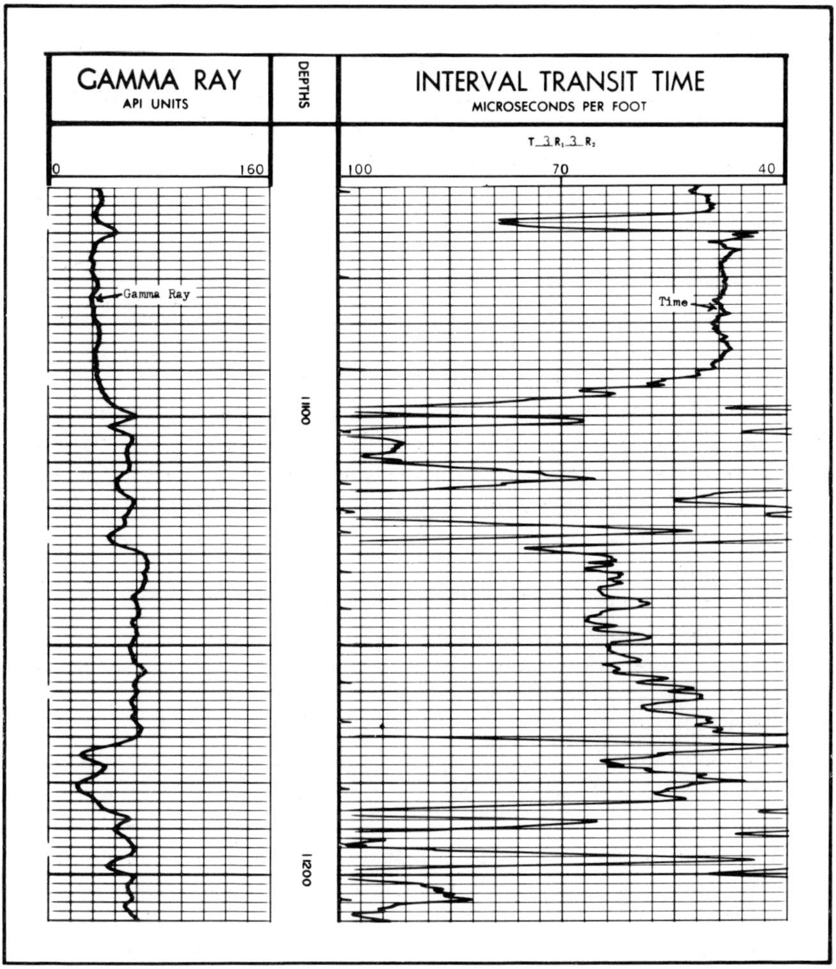

Sonic or Acoustic Log
Figure 37

travel times are:

1-	Dolomite	23,000 feet per second	43.5 microseconds
2-	Limestone	21,000 to 23,000 Ft/second	47.5 - 43.5 microseconds
3-	Sandstone	18,000 to 19,500 Ft/second	55.5 - 51 microseconds
4-	Shales	varies up and down around 10,000 Ft/second	100 ± microseconds

Velocity through water is around 5300 feet per second, which is 189 microseconds.

As the rock becomes more porous the sound travels more slowly, causing the transit time to increase. Therefore, the higher the transit time the greater the porosity. This holds true only when the rock is shale-free and there is liquid in the pore spaces.

The transit time through a formation will be increased if the formation becomes shaley. Gas in the pore spaces also increases the transit time. The transit time may be affected in unconsolidated or uncompacted sandstones, as well. This generally causes an optimistic porosity value when using the Wyllie relationship. A compaction correction is necessary to derive proper porosity values.

The transit time from the Sonic Log may be integrated and used by Geophysicists in interpreting seismic data.

A Caliper Log and Gamma Ray Log may be run with the Sonic Log.

§ 201(b) Density Log

The next approach to porosity measurement was the Density Log, a measurement of the electron density in a formation, which varies with porosity.

An instrument that will measure formation density is shown in Figure 38. This instrument is mechanically pressed on the wall of the hole so that the measurements are made as directly on the rocks as possible. In the portion of the tool that fits closest to the hole wall is a source of gamma rays and two detectors. The instrument with two detectors produces the Compensated Density Log (FDC). The old density logging equipment had one detector

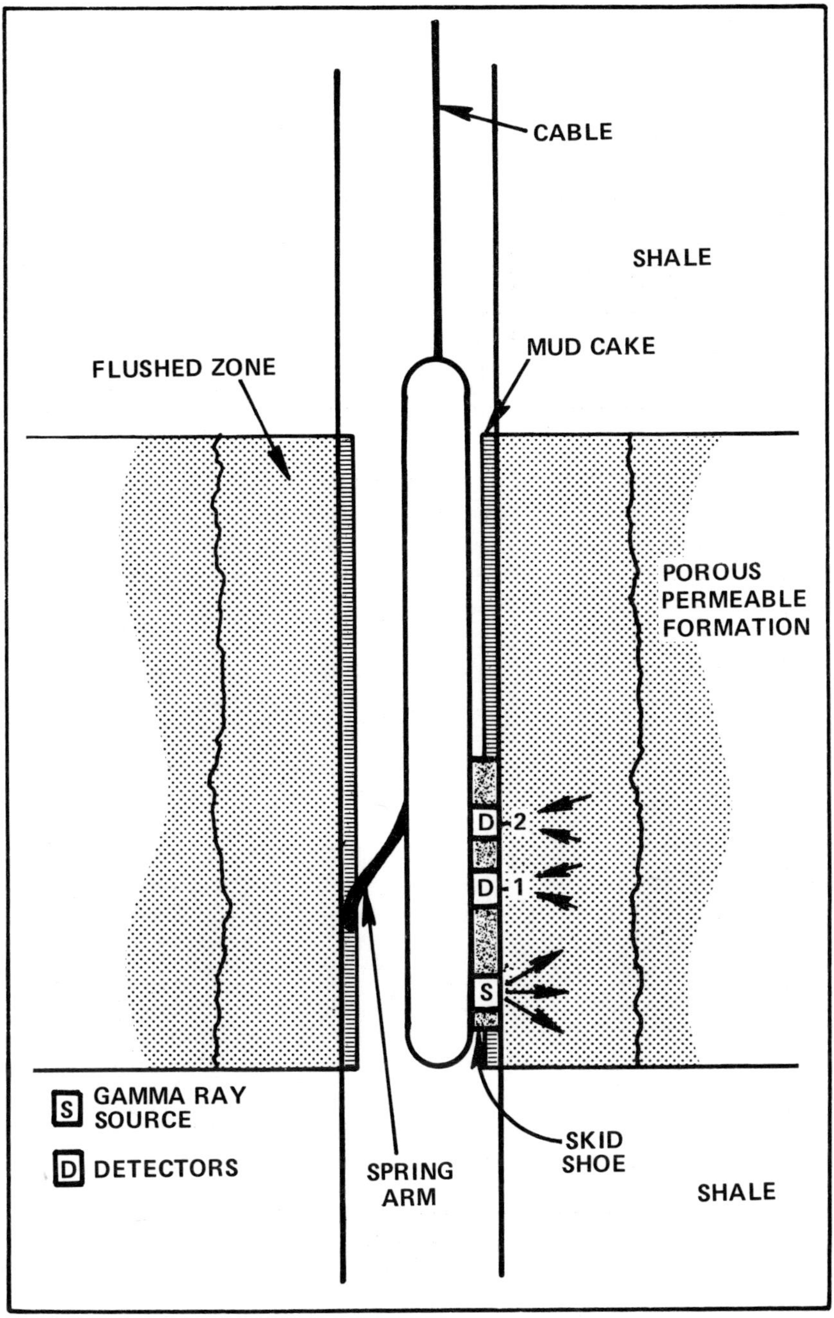
CABLE
SHALE
MUD CAKE
FLUSHED ZONE
POROUS
PERMEABLE
FORMATION
D-2
D-1
S
S GAMMA RAY SOURCE
D DETECTORS
SPRING ARM
SKID SHOE
SHALE

Figure 38

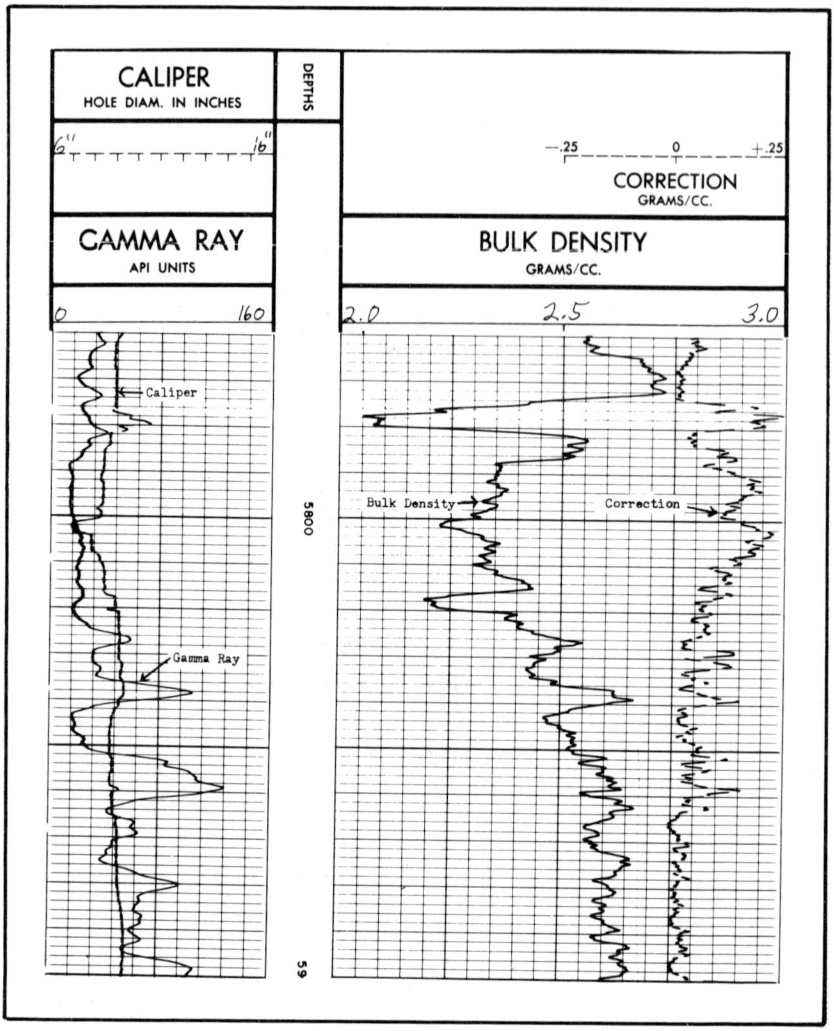

Density Log
Figure 39

which permitted measurements that could have adverse effects from the mud cake and irregular hole sizes. The gamma rays emitted are of such an energy level that they can be absorbed by the electrons in the formation. The source and the detectors are shielded in such a manner that the only opening is in direct contact with the rock. The gamma rays leave the source and go directly into the formation. If the formation is dense (no porosity), the gamma rays are absorbed very close to the source, and very few are counted by the detectors. As porosity increases the formation becomes less dense and gamma ray absorbtion is decreased. More gamma rays reach the detectors and the counting rate is increased.

The electron density is related to the weight of the rock in grams per cubic centimeter. Most of the rocks encountered in a well will weigh between two and three grams/cc; therefore, the log is scaled to accommodate these values. The Density Log is recorded in tracks 2 and 3 of the grid and a typical log is shown in Figure 39.

The Compensation Curve at the right side of the log is a recording of the amount of correction the equipment made on the Bulk Density Curve. This is useful in that it allows the log user to determine whether or not the Bulk Density Curve readings are correct.

The control equipment also will solve the Bulk Density Curve for porosity and record this value simultaneously on the log. A grain density must be chosen by the engineer running the log, who generally knows the geologic column and will use grain density values appropriate to the area.

The Density Log investigates a volume of rock very close to the wall of the hole so that in a porous permeable formation it will measure predominately in the flushed zone. Consequently, it will include the fluid (mud filtrate) in its measurement. The density of that fluid must be considered when the Density Log is solved for porosity. The relationship that permits this solution is:

$$\emptyset = \frac{\rho\ \text{grain} - \rho\ \text{log}}{\rho\ \text{grain} - \rho\ \text{fluid}}$$

Where: ρ grain = weight of rock at "0" porosity
ρ log = weight as measured by the log
ρ fluid = weight of fluid in the porous space

Some common weights are:

1- Dolomite = 2.87 grams per cubic centimeter
2- Limestone = 2.71 grams/cc.
3- Sandstone = 2.65 grams/cc.
4- Shale = 2.2 to 2.65 grams/cc. Generally lower in shallow or noncompacted shales - higher in deeper ones.
5- Fresh water = 1.0 grams/cc.
6- Salt water = 1.1 grams/cc.

If gas is present in the pore spaces it will have an effect on the measurements. The gas effect is to cause the porosity value determined with the above relationship to be greater than it should be, since the density of gas is less than the density of water.

Because the heavier shales weigh near that of the reservoir rocks, they will have a lesser apparent effect on the porosity values determined from the Density Log than those from the Sonic Log.

The mechanical arrangement that presses the density instrument against the wall of the hole follows the contour of the hole size changes. Taking advantage of this allows a caliper curve to be run with the Density Log. A Gamma Ray Log may also be run with the Density Log.

Improvements to the neutron logging technique came along next, improvements which made modern neutron logging instruments provide very accurate porosity values.

§ 2.01(c) Neutron Logging

Neutron Logging is a measurement that indicates the presence of hydrogen in water or oil. Water and/or oil occur in the porous spaces, therefore the Neutron Log is a porosity sensitive log.

The basic principle of neutron logging may be illustrated in the following manner. Consider a cue ball used in billiards and the effect on other billiard balls when hit by the cue ball. If the cue ball hits another ball at an angle it imparts part of its energy to that other ball and both go in different directions. If the cue ball hits another ball head-on it imparts all of its energy to the other ball. The cue ball stops dead and the other ball moves off. If the cue ball were to hit a bowling ball, whether at an angle or head-on, it imparts practically no energy to the bowling ball because it is so much heavier than the cue ball. Neutrons will behave in a manner similar to that of the cue ball, imparting some or all of its energy to another atom of equal weight. The atom of near equal weight is the hydrogen atom.

Nearly all neutron logging devices have some common characteristics. They have a source of neutrons and one or more detectors.

The source emits neutrons that bombard the formation. If one should strike a hydrogen atom and impart its energy to that atom very high energy "gamma rays of capture" will be given off. The detectors in the equipment are designed to measure these gamma rays of capture. The Neutron Log, then, responds to the presence of hydrogen.

Many neutron logging devices have sources that emit neutrons of an energy level that the log responses may be affected by many things such as mineral make up of the rocks, bore hole fluid, bore hole size, and tool position in the hole, to name only a few. An example of such an instrument is shown in Figure 40.

Important improvements in the neutron logging technique have been made in recent years, and one of them is the Sidewall Neutron Porosity Log. The instrument which produces this log is illustrated in Figure 41.

This instrument is mechanically pressed against the wall of the hole, the source and detector are mounted in the skid shoe so that they are in as

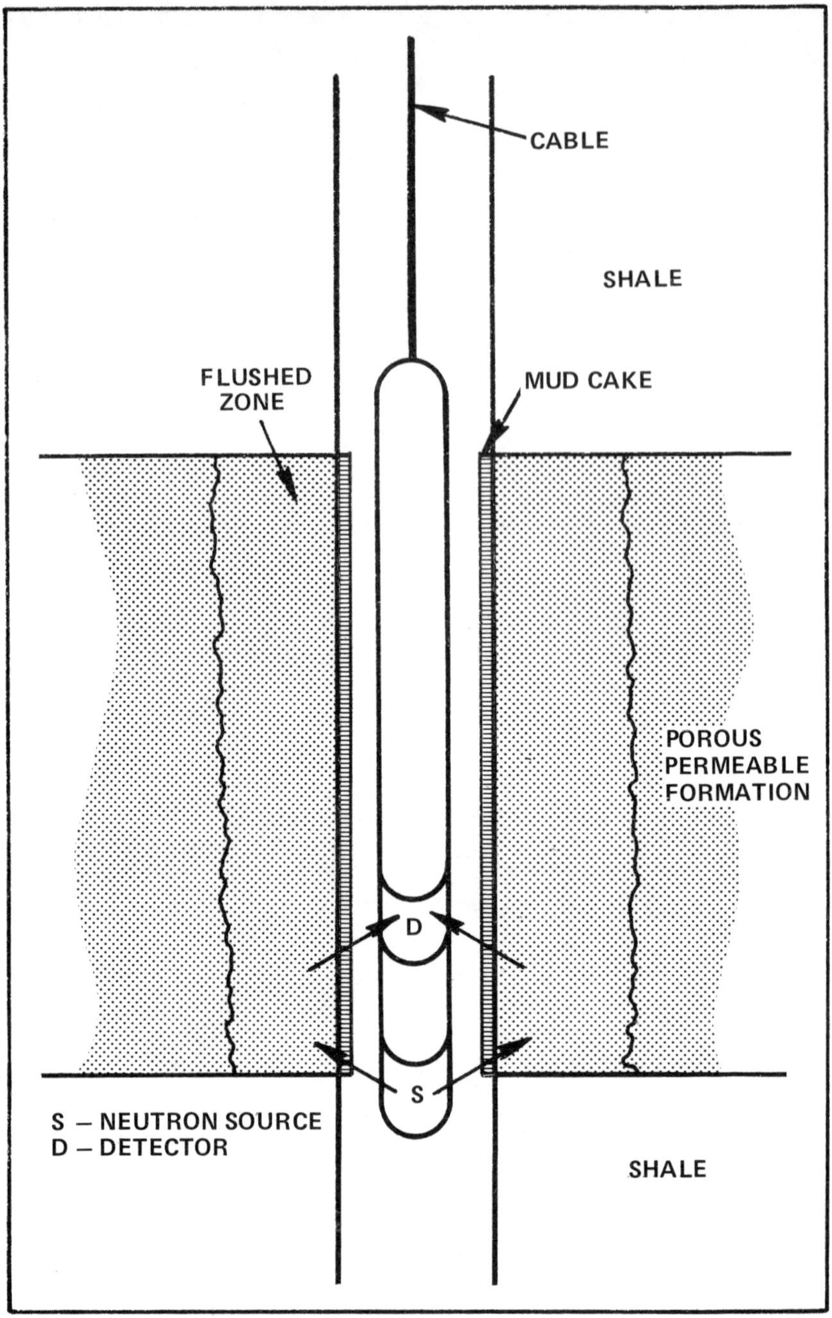
CABLE
SHALE
FLUSHED
ZONE
MUD CAKE
POROUS
PERMEABLE
FORMATION
D
S
S – NEUTRON SOURCE
D – DETECTOR
SHALE

Figure 40

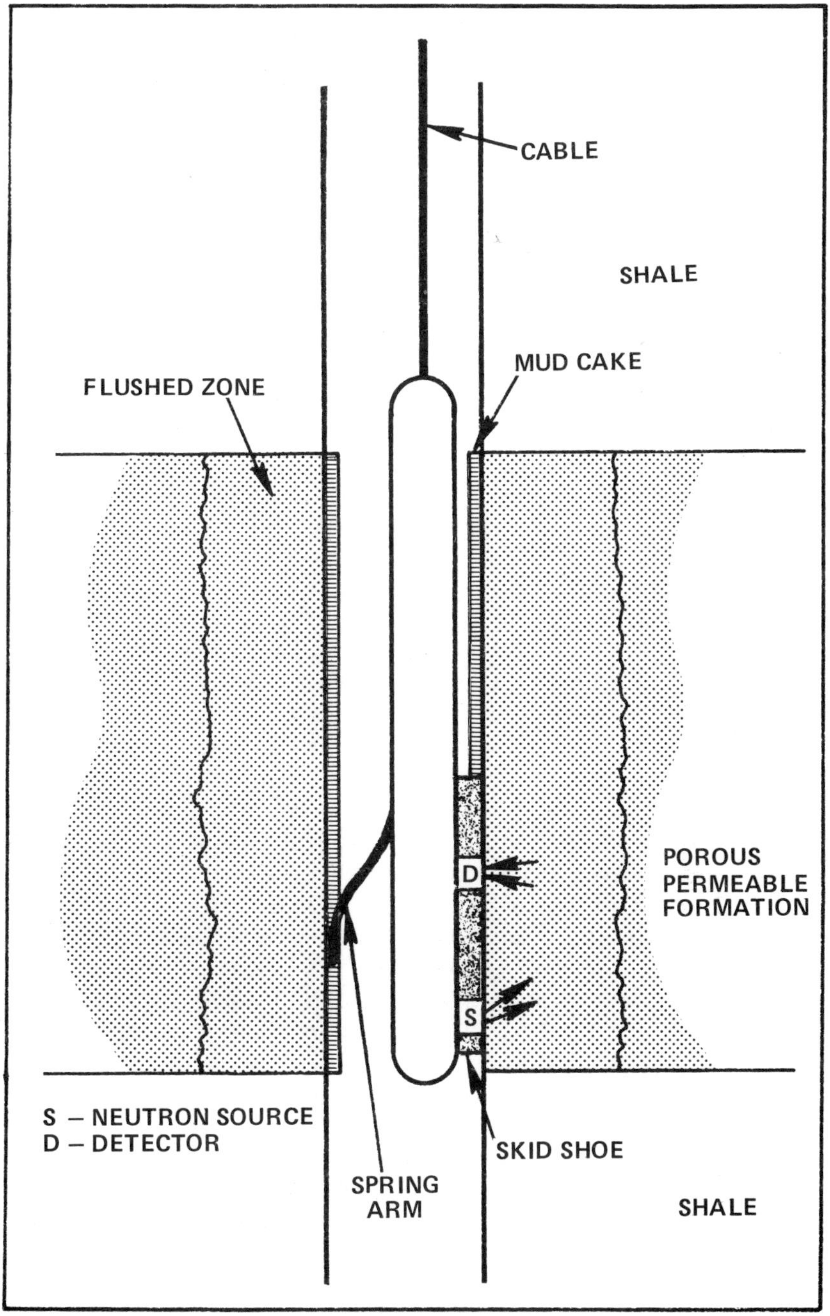
CABLE
SHALE
MUD CAKE
FLUSHED ZONE
D
POROUS
PERMEABLE
FORMATION
S
S – NEUTRON SOURCE
D – DETECTOR
SKID SHOE
SPRING
ARM
SHALE

Figure 41

direct a contact with the rocks as possible. The source emits neutrons of a much higher energy level. These improvements reduce and almost eliminate the effects of mineral mixture, bore hole fluids and size, so long as the instrument remains in contact with the hole wall. The control equipment makes whatever corrections are necessary. This log makes its measurements very close to the bore hole and will be predominately in the flushed zone of porous permeable rocks. It may have a caliper and gamma ray run with it, also.

The Sidewall Neutron Porosity Log (SNP) may be run in any bore hole environment, whether filled with fluid or empty, and only in uncased holes. An example of this log is shown in Figure 42.

Further improvements produced the Compensated Neutron Log (CNL), which is a mandrel type instrument with a source of neutrons and two detectors. Such an instrument is shown in Figure 43.

The Compensated Neutron Logging method uses a little different technique to accomplish its measurements. The detectors are designed to measure neutrons of the thermal energy level, rather than "capture gamma rays". The emitted neutrons are of a higher energy level than thermal, and when they collide with hydrogen they lose energy and are slowed down. When they are slowed to the thermal level they are detected by the detectors. The end results are the same, though, which is the detection of hydrogen. The control equipment uses the ratio of the two detector counting rates to practically eliminate the bore hole effects. The resulting measurements provide values that are presented in tracks 2 and 3 on the log as a linear porosity scale. Figure 44 is an example of the Compensated Neutron Log.

This log has a deeper investigation into the formation than the Sidewall Neutron Log, a feature that is useful in some interpretation techniques such as detection of gas. This will be explained in a later section.

The Compensated Neutron Log is designed to be run in either cased or uncased holes, and only in those that are liquid filled. It cannot be run in empty holes.

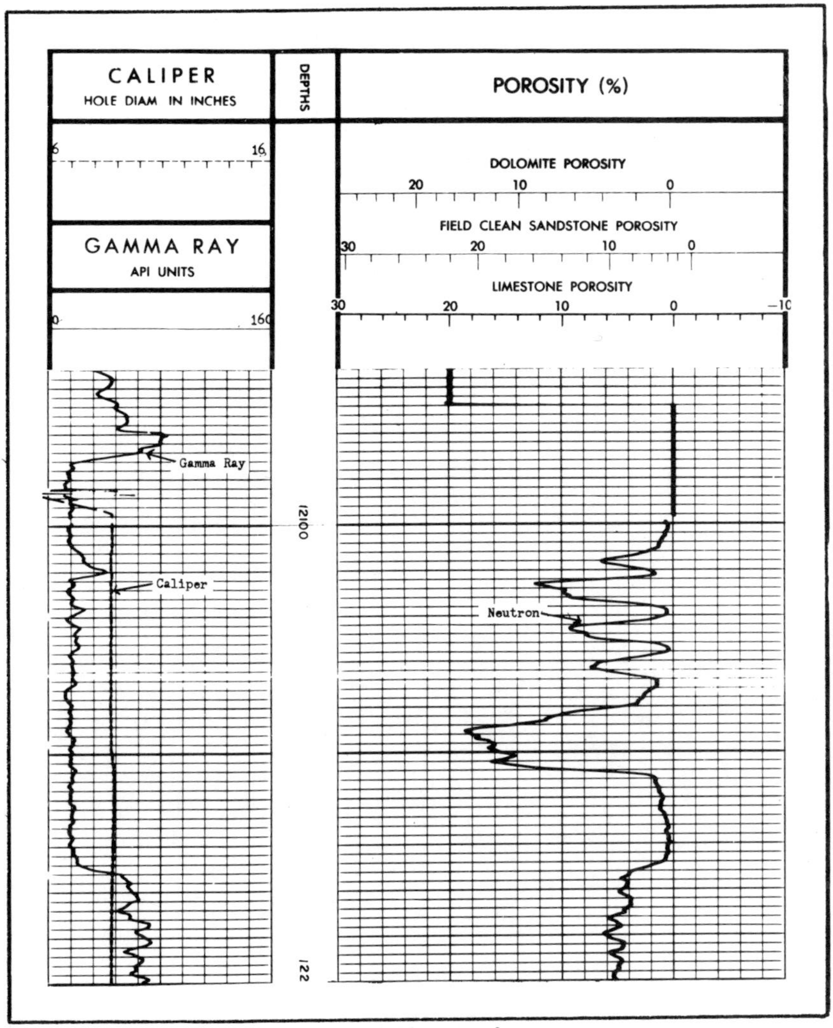

Sidewall Neutron Log
Figure 42

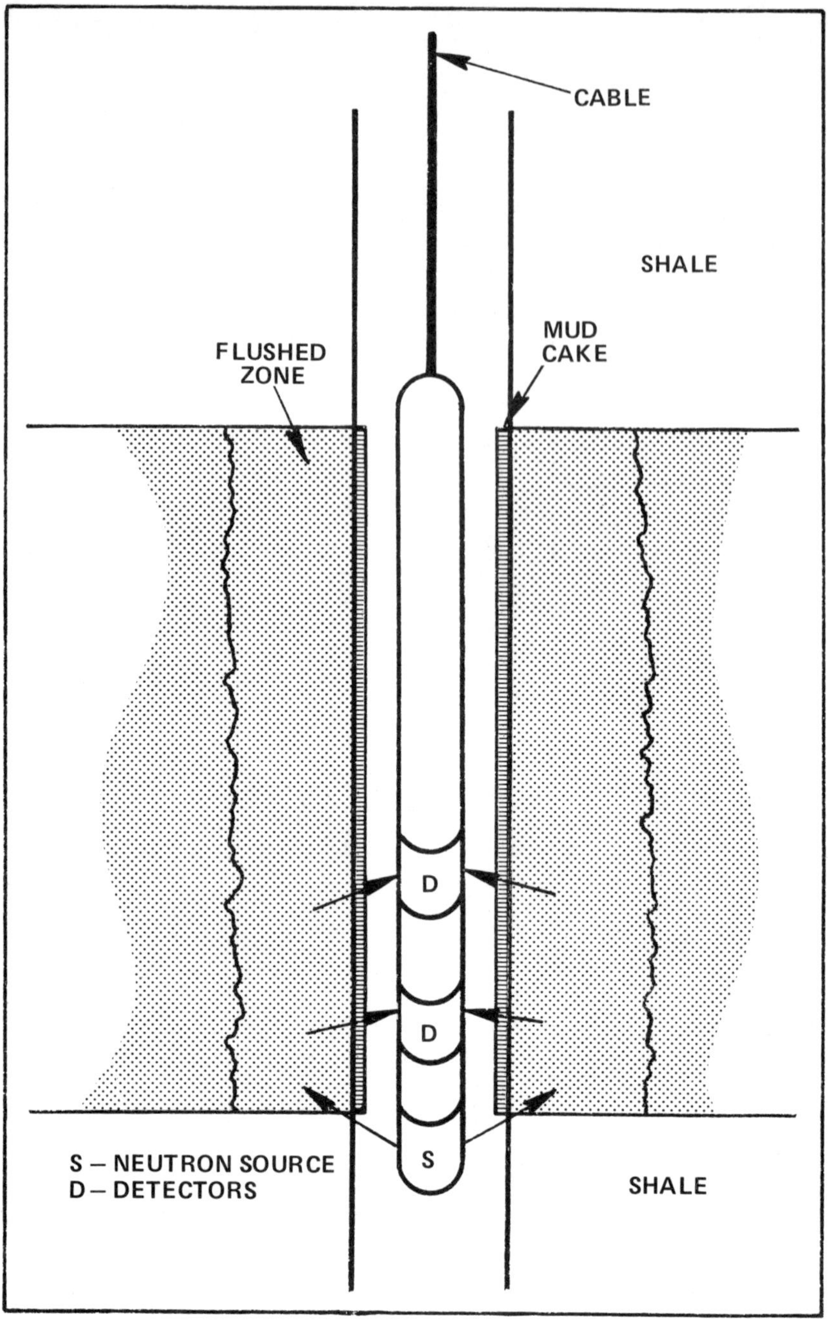
CABLE
SHALE
MUD
CAKE
FLUSHED
ZONE
D
D
S
S – NEUTRON SOURCE
D – DETECTORS
SHALE

Figure 43

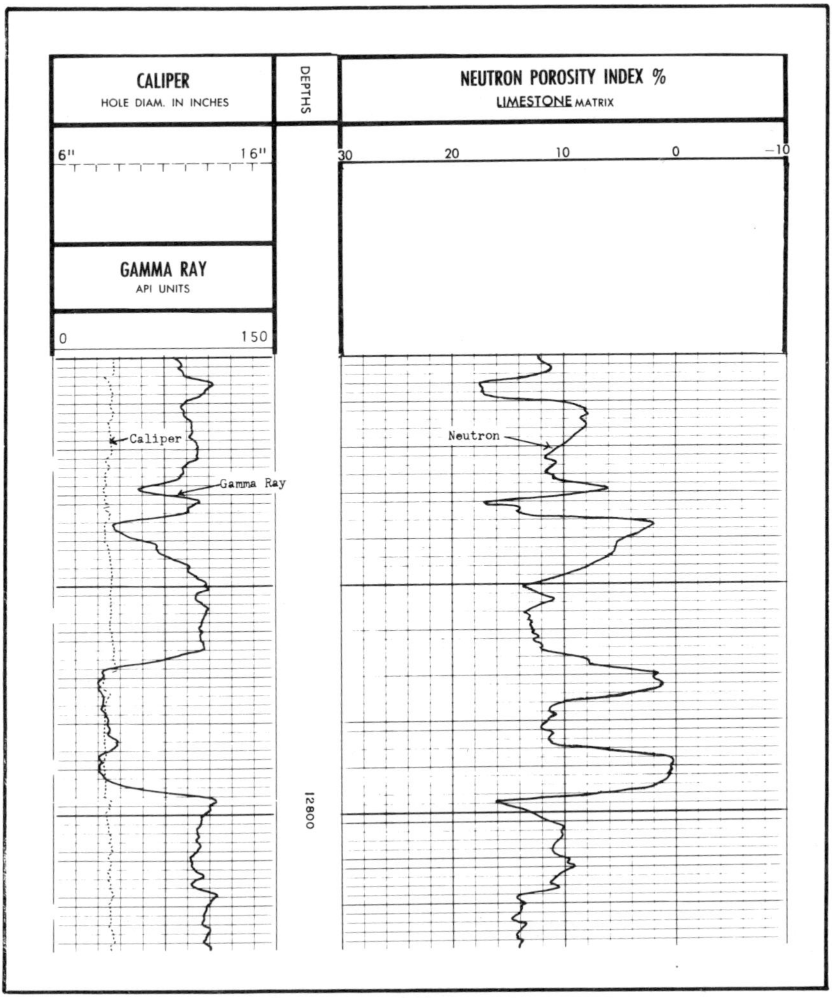

Compensated Neutron Log
Figure 44

As with other Neutron Logs, the Gamma Ray Log can be run with this one. In cased holes a casing collar log can be included.

All Neutron Logs have one thing in common: they detect the presence of hydrogen in the liquid. Should there be dry gas in the formation there is much less liquid, so there are fewer hydrogen atoms to be detected. Consequently, the porosity indications are decreased. The Neutron Logs then will be affected by gas in a pessimistic manner. The Neutron Logs will respond to the presence of hydrogen in the water contained in shale, so shale will cause the Neutron log to indicate porosity. The Gamma Ray will provide information to identify the shales.

Let us recap the porosity logs before we move on to the next subject.

The microdevices are instruments that measure resistivity and are primarily limited to measurements in the flushed zone. From these measurements F is determined, and from F porosity is estimated. These measurements are affected by the mud filtrate resistivity, the mud cake resistivity, the mud cake thickness, and residual hydrocarbon saturation. Any one or all of these variables may or may not be determined with any degree of accuracy. Therefore, porosity determination from these logs may not be very accurate. Resistivity measurements made with these devices, especially the focused types, are used in various interpretation techniques which will be discussed in the section on interpretation.

The Sonic or Acoustic Log is a measurement of the time it takes sound to travel through the rocks. It is affected by the mineral mixture of the formations, the liquid (or gas) in the porous spaces, and shale. Gas in the porous spaces and shale in the formation causes the travel times to increase. This makes porosity derived from sound measurements to be higher than it should be.

The Density Log is a measurement of the electron density of the rocks, which is related to porosity. It is affected by the mineral mixture of the formations, the liquid (or gas) in the porous spaces, and, to a lesser extent, by shale. Gas in the porous spaces has less density than liquid, so that porosity derived from density measurements is higher than it should be.

Since shale and sandstone densities are generally similar, porosities in shaley sands will be more accurate. However, there may be an optimistic effect in shaley carbonates.

The Neutron Logs measure the presence of hydrogen which is generally in the liquids in the porous spaces. They are affected by the mineral mixtures of the formations, the liquid (or gas) in the porous spaces, and shale. Gas in the porous spaces has less liquid so there is less hydrogen. The porosity derived from neutron measurements is lower than it should be when gas is involved. The neutron will respond to the hydrogen in the water in the shale so that shaliness can cause the porosity to be higher than it should be.

From this we can see that gas has an optimistic effect on two of the porosity logs, Acoustic and Density, and a pessimistic effect on one, the Neutron. Shale has an optimistic effect on two, Acoustic and Neutron, and a lesser effect on one, the Density. All three are affected by the mineral mixture of the formation.

In order to turn these adverse effects to our advantage, these logs are often run in combination. When two (or all three) are run together interpretation techniques are used which aid in the determination of the mineral mixtures that make up the various formations, and detect the presence of gas in the porous spaces and the amount of shale present in the formations. The techniques will be discussed in the interpretation section.

Chapter Three

AUXILIARY LOGGING TECHNIQUES

The logs we have discussed up to this point are considered essential, and provide much of the information needed to make decisions as to whether or not the well should be completed. There are other logs available that can provide additional information that may be necessary, depending upon the classification of the well.

Some of these logs are designed to be run in uncased holes, while others are designed for use in cased holes. Still more logs may be used in either uncased or cased holes.

Some of these auxiliary logs are:

Uncased holes	Cased holes	Both
Dipmeter Log - Diplog	Production Logs	Temperature Log
Fracture Identification Log	Thermal Decay Time Log	Amplitude Log
	Neutron Lifetime Log	Variable Density Log

§ 3.01 The Dipmeter Log - Diplog

Modern dipmeter logging equipment (See Figure 45) makes four simultaneous resistivity measurements on the wall of the hole. The electrode systems are located in pads on four spring arms 90° apart and all in the same plane. The measurements are focused microresistivity measurements, so that very great detail is obtained. The opposite arms are interlocked in such a fashion that two calipers also are recorded at 90°

phasing. Other measurements made at the same time are the tool position in the hole, and the azimuth of the #1 electrode.

The values measured are recorded in digital form on magnetic tape, and recorded at the same time in analog form on paper or film. The digital magnetic tape is in a form that can be used with computers to compute the dip angle and direction of formation bedding planes. The measurements are in such great detail that manual computation is not practical. An illustration of a typical four-arm dipmeter analog presentation is shown in Figure 46.

The results of the computation are shown in Figure 47. The amount of dip and direction are plotted as a series of "tadpoles", with the head being the angle, and the tail being the direction, of the dip. At the extreme right are more "tadpoles" that show, and approximate, the tool drift and direction. In track 1 (left of the depth column) is a correlation curve. Sometimes it is a resistivity curve, and sometimes it is a spontaneous potential curve. This curve is correlated with the open hole log, so that exact depth intervals may be studied.

Geological happenings such as faults, unconformities, sand bars, channel fill sands, reefs, folds, and others, generate typical patterns on the "tadpole" presentation. Also, formation geometry, direction of fluid flow, and depositional environments can cause patterns to form. Figure 48 indicates some of these patterns.

From these patterns experienced analysts can determine the various anomalies. It is a very complex process and should not be attempted by a novice.

Of course the dip patterns will determine the direction to go to gain structural advantage, such as in the offset well, or to get up dip from a water-filled zone.

The two calipers can give good information on the hole geometry for testing, cementing, or other uses.

§ 3.02 The Fracture Identification Log

The four arm dipmeter equipment is used to run the fracture

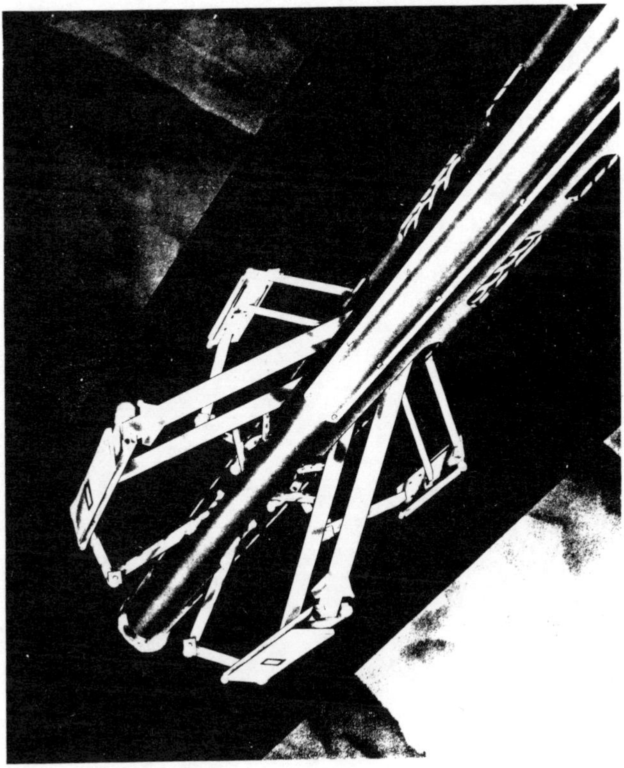

4 Arm Dipmeter Logging Tool
Figure 45

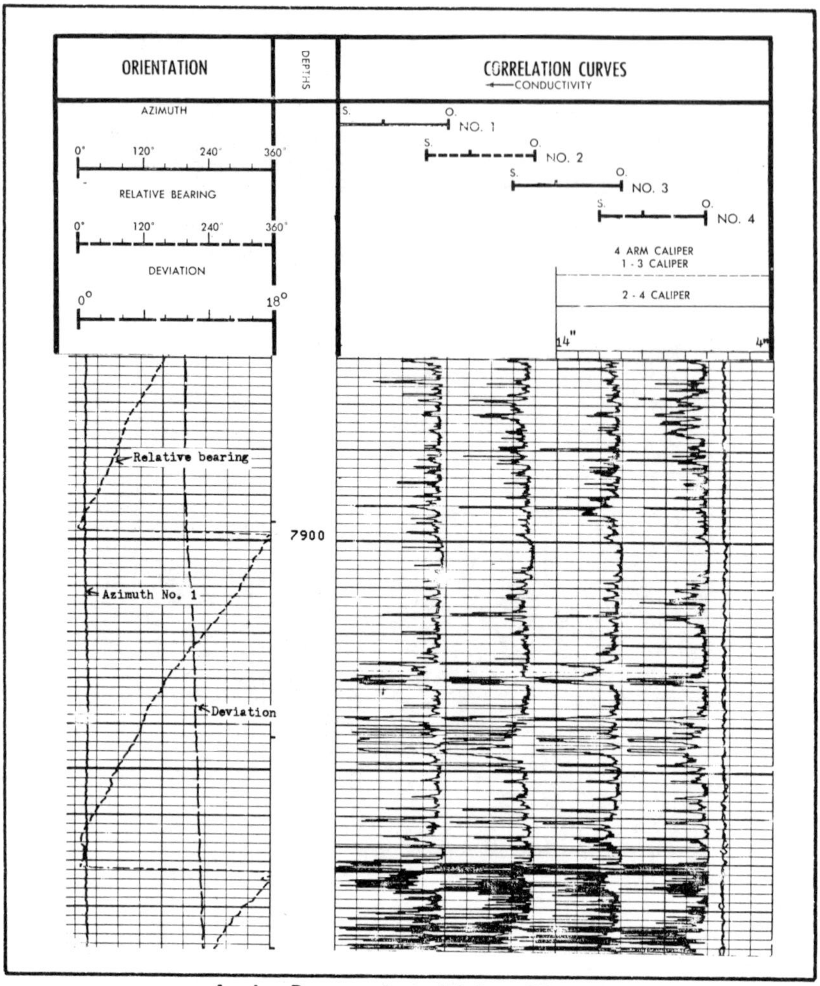

Analog Presentation of 4 Arm Dipmeter
Figure 46

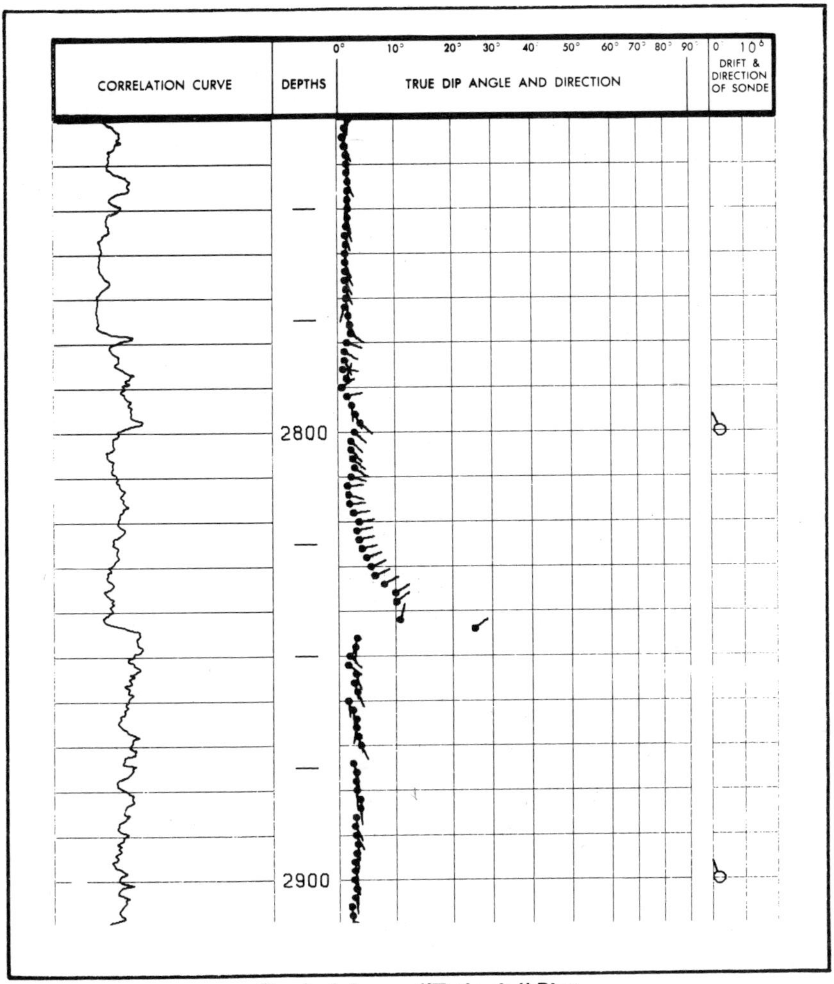

Typical Arrow "Tadpole" Plot
Figure 47

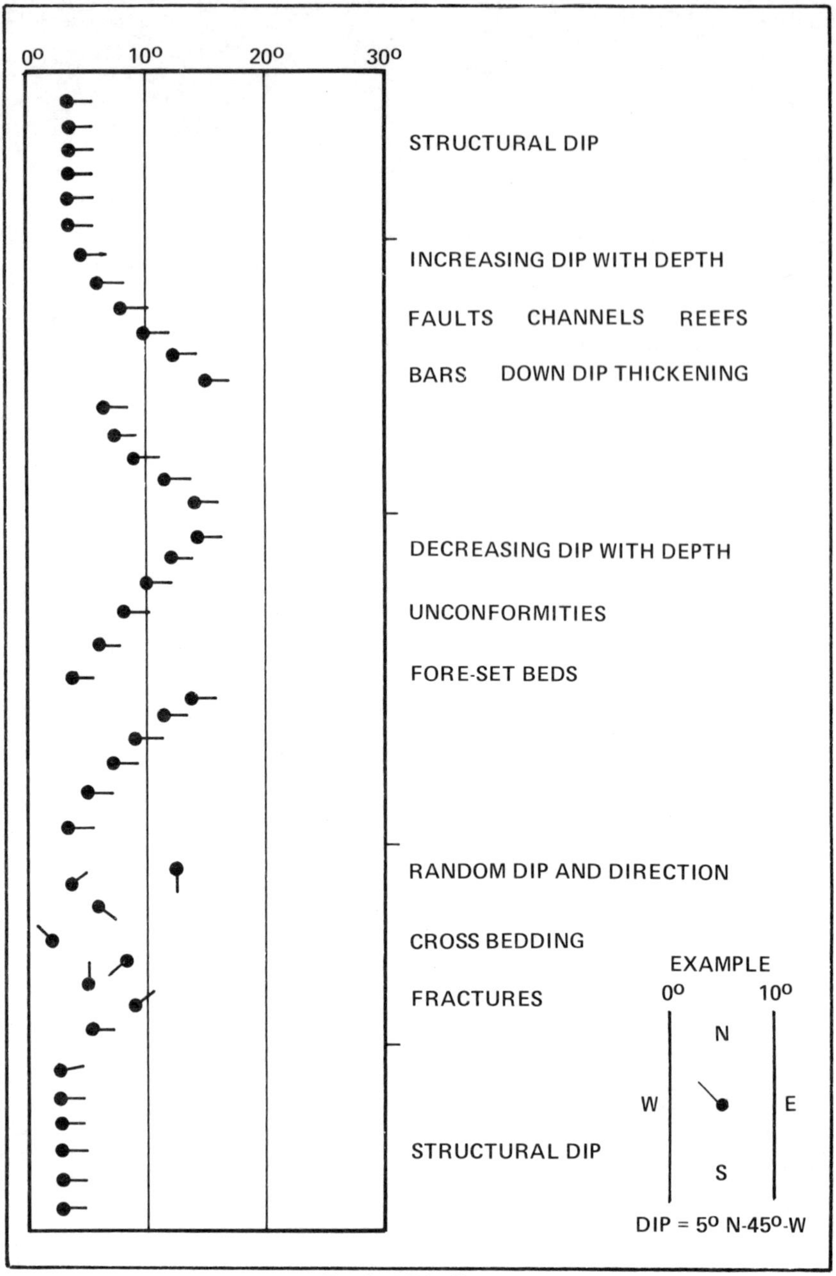

Typical Dip Patterns
Figure 48

identification log. The presentation of the measurements is somewhat different than for the dipmeter. Only the analog record on film is made, since the analysis of the log is visual and not by computer. Figure 49 is an example of the fracture identification log, and Figure 50 is a schematic illustration of the principle of operation. If a fracture crosses the bore hole at or near the center, it may affect the resistivity as measured by opposite pads. If the fracture is open it will fill with mud. This causes a decrease of resistivity as compared to the unfractured rock. The resistivity measured by the two opposite pads on the fracture will be less than the resistivity measured by the two pads not on the fracture. The fracture may cross the hole off-center and may affect only one measurement at a time, or two adjoining arms rather than opposite ones. The fractures may be "healed", or filled with some sort of solid material. Whether or not these fractures affect the measurements will depend upon the filling material and whether or not water is in the fracture. If the resulting resistivity is less than the original matrix resistivity, the fracture will show on the log. If the resistivities are similar, the fracture may not show.

The direction of the fracture trend generally can be determined from the orientation of the # 1 electrode arm. If the fracture effect is on # 1 and # 3 electrode arms, then the direction will be the same as # 1 orientation. If the fracture effect is on # 2 and # 4 electrode arms, the direction will be # 1 orientation plus 90°. The fracture may not cross the hole at the center, but be in a position that causes it to affect adjoining electrode arms. In this case, the direction of the fracture must be calculated using the orientation of #1 electrode arm and the affected electrode arms. If the fracture does not cross the hole, or if it affects one electrode arm and not the others, the direction must be calculated in this instance also. Figure 51 illustrates these procedures.

This tool will rotate in the hole as the log is run, so repeat runs are generally made so that confirmation of the fractures and directions may be made. This is possible because, due to the rotation, the fracture effect may be seen on different sets of measurements.

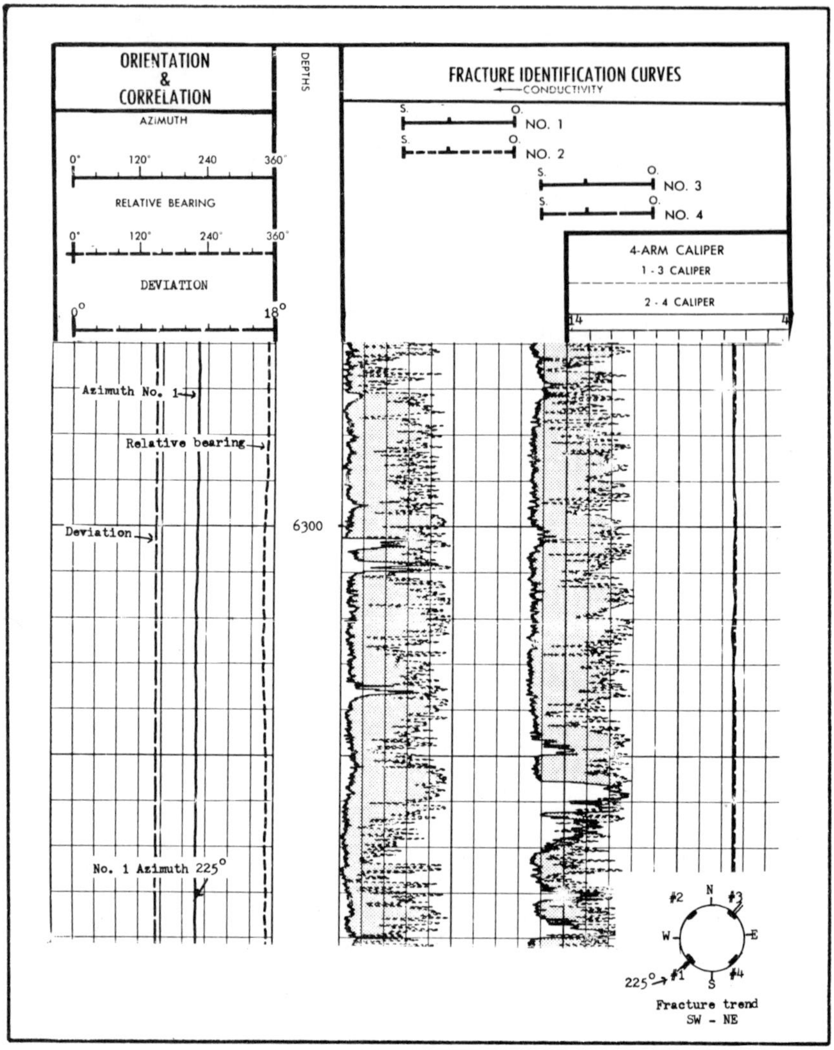

Fracture Identification Log
Figure 49

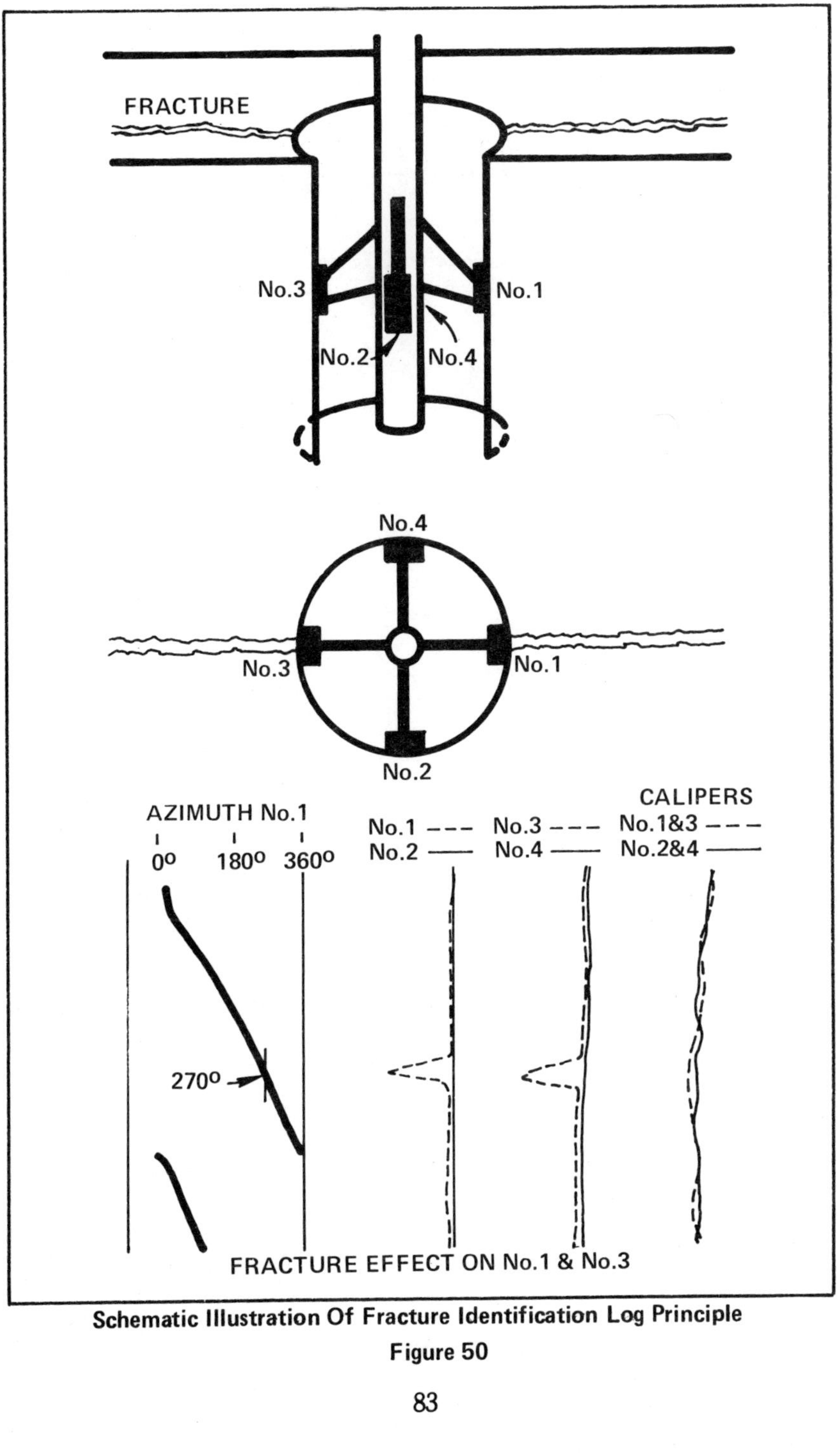

Schematic Illustration Of Fracture Identification Log Principle

Figure 50

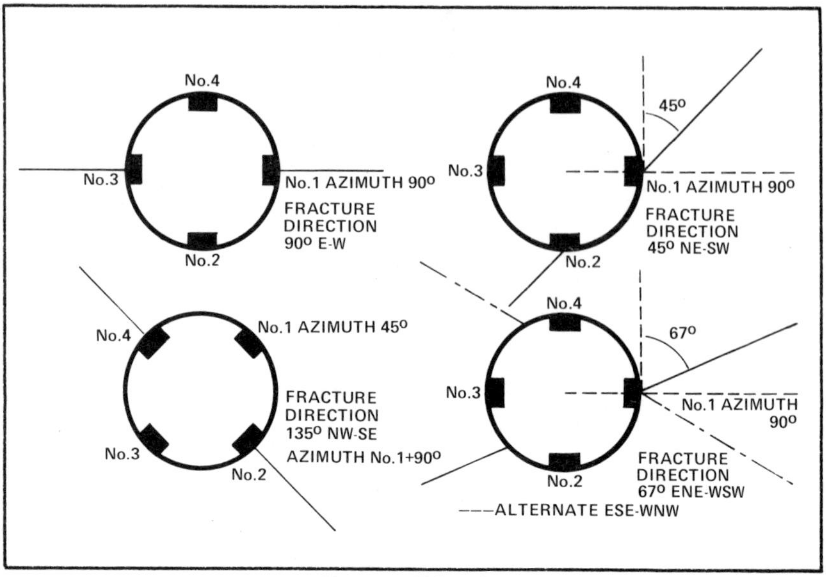

Fracture Direction
Figure 51

The dual caliper often helps because the hole can enlarge in the direction of the fractures, yet remain in gauge in the other direction.

The Fracture Identification Log usually is recorded on a film scale of 24″ equals 100′ of depth. This permits a very detailed study to be made.

The identification of fractures may be very helpful when completing the well, such as selection of perforation density and intervals to be perforated, and selection of stimulation techniques.

§ 3.03 The Temperature Log

The temperature log may be used in either uncased or cased holes, serving different purposes in each case. In the uncased hole it is used to locate gas-producing zones in a well drilled with air or gas. It also is used to help locate zones of lost circulation in a well drilled with mud.

In cased holes the temperature log is used to locate the top of the cement fill-up behind the pipe.

These uses take advantage of the fact that the temperature in the earth increases with depth. The rate of increase is called the temperature gradient and may vary in different geological areas.

§ 3.03(a) Location of Gas-Producing Zones. When an air or gas-drilled well penetrates gas zones the pressure is released and gas is free to flow into the wellbore. This flow causes expansion of the gas and produces a cooling effect, generally upward from the entry point.

A temperature log run into the hole will indicate a normally-increasing temperature until the instrument becomes influenced by the cooling of the expanding gas. Cooling will continue until the instrument is below the producing zone. Temperature will then attempt to reach the normal gradient again. Should there be another gas-producing zone below the first one, there will be another cooling effect. The shape of the curve will depend on the distance between the two zones, as illustrated in Figure 52. There is a technique that permits an approximation of the gas flow from these zones.

§ 3.03(b) Location of Lost Circulation Zone. Sometimes the drilling mud no longer returns to the surface as it should in the normal drilling process. This is usually caused by the mud being lost into some subsurface formation or unusual condition, and the mud being lost is continually replaced by new mud from the surface pits. This can cause the mud column above the lost circulation zone to be somewhat cooler than the mud column below the lost circulation zone, which has been static and tends to reach formation temperature gradient. There should be a discontinuity of the temperature gradient near the zone of lost circulation. Such an indication is shown in Figure 53.

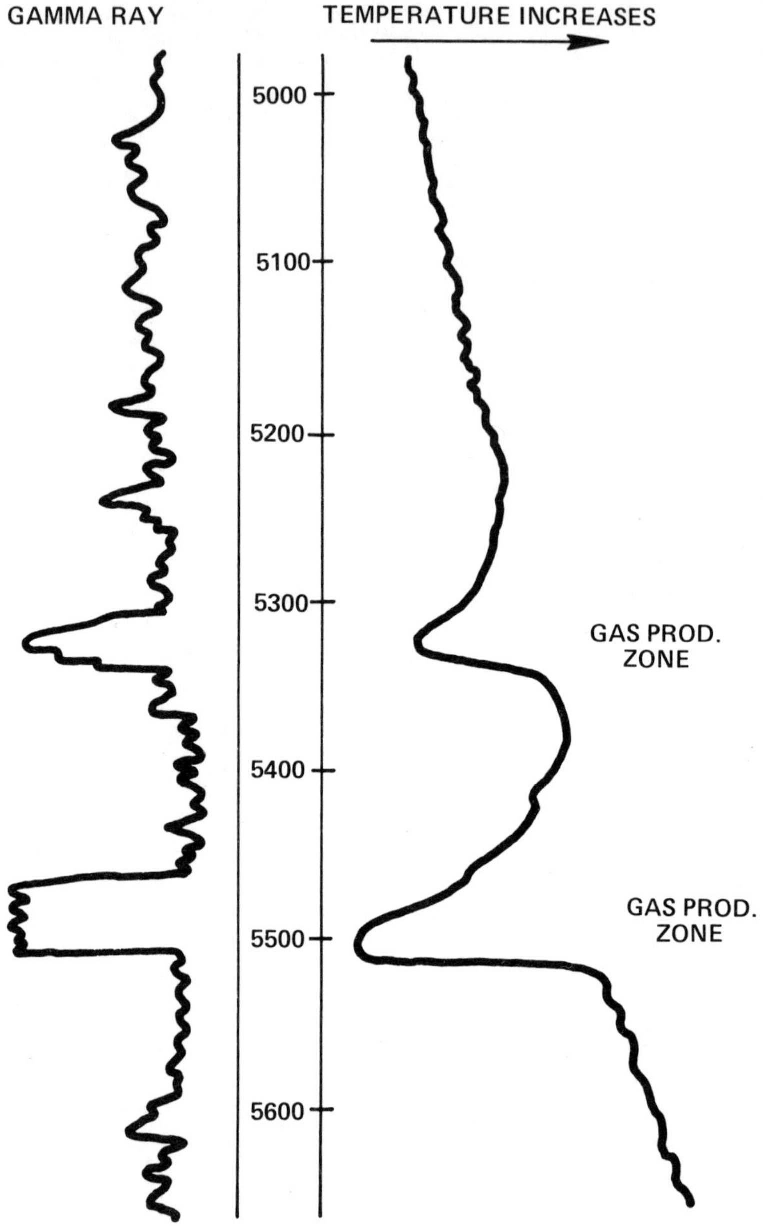

Temperature Log In Gas Producing Zones—Empty Hole

Figure 52

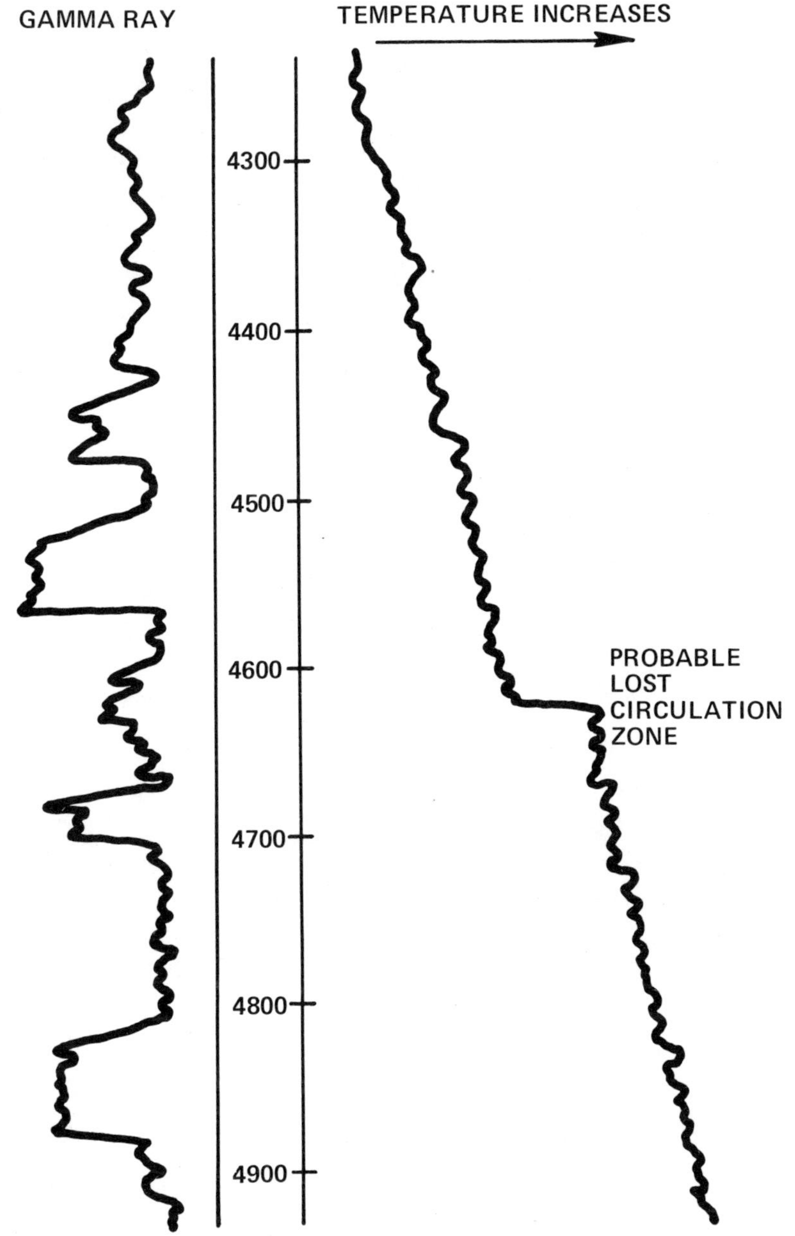

Temperature Log in Lost Circulation Zone

Figure 53

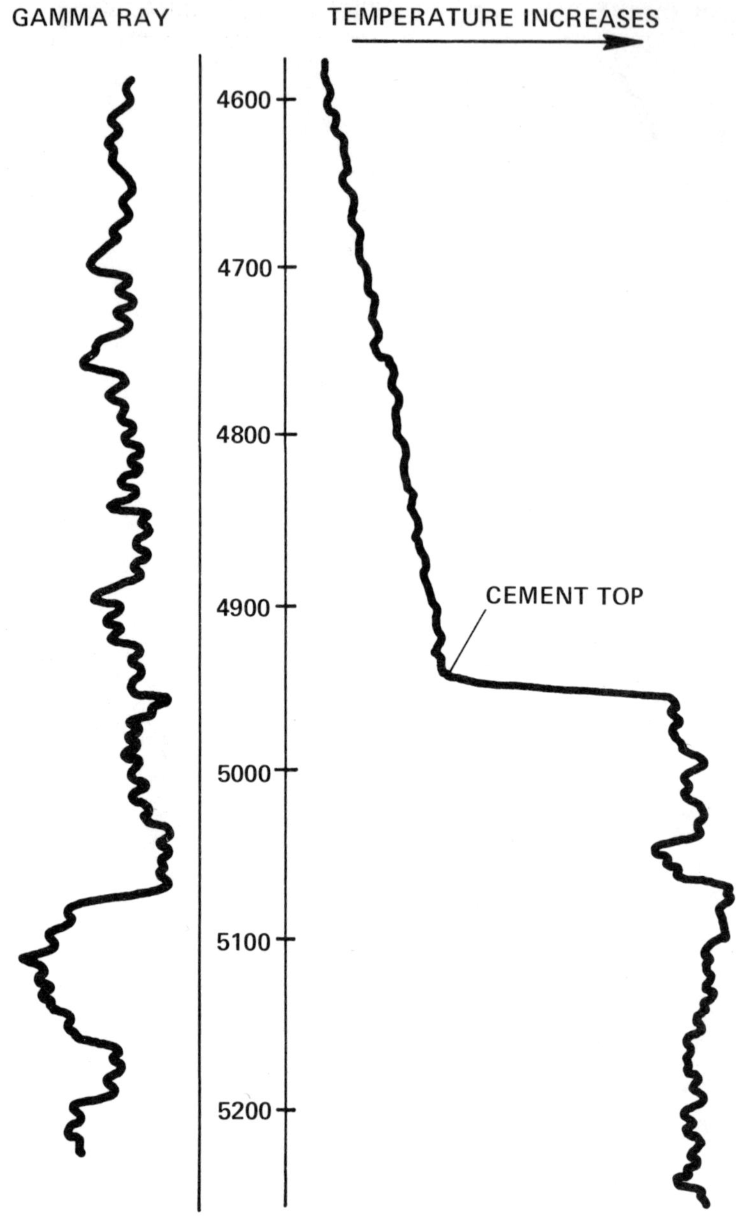

Temperature Log—Top Of Cement

Figure 54

§ 303(c) Location of Cement Top. When casing is set in a bore hole, cement is pumped down the inside of the pipe, out the bottom, and up into the annular space between the casing and the hole wall. It is necessary to know how far up the top of the cement is, to assure proper protection and segregation of the zones of interest.

When cement sets it generates considerable heat for some time. If a temperature log is run during this time it will measure this increase and indicate where the top of the cement is. For best results the log should be run within about 12 hours. An example of cement top location is shown in Figure 54.

The Temperature Log does not tell much about the quality of the cement job, so far as channels, voids, and mass.

Another type of temperature log that has some special applications is the Differential Temperature Log, which is a continuous measurement of the difference in temperature between two subsurface points. Temperature differences will occur where there is some fluid or gas movement. Because of this, the differential temperature log is used in flowing, pumping, input or injection wells. Its uses include location of channels behind pipe, either gas or water, gas/water contacts, water producing zones, and injection zones in enhanced recovery wells, among others.

§ 3.04 The Amplitude Log and The Variable Density Log (Micro-Seismogram Log)

These logs may also be used in either uncased or cased holes. In uncased holes they are used as an aid to detect fracture systems, and in cased holes they are used as a cement bond evaluation log.

The Sonic or Acoustic logging equipment is used to run these logs. However, rather than time measurements, the amplitude or energy level of the sound wave is measured. When a sound pulse is transmitted it is in the form of a sine wave and will have a given amplitude, as shown in Figure 55.

If this amplitude is measured, and nothing happens to cause a loss of energy, the measurement will remain near full value.

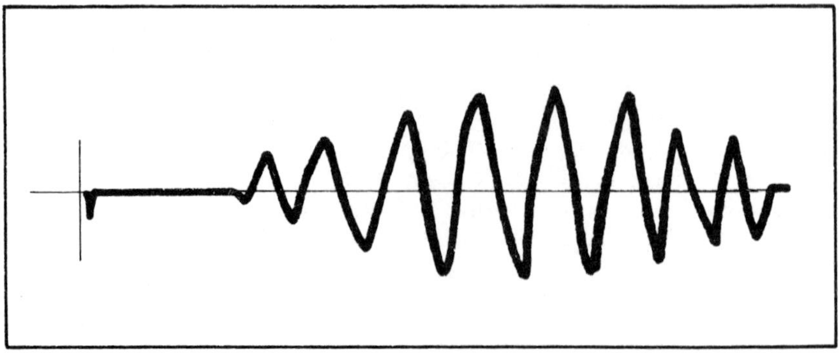

Sound Pulse Sine Wave
Figure 55

§ **3.04(a) Uncased Hole Application.** The usual use of the amplitude log in uncased holes is to detect fractures in formations. If the rock is massive it will transmit the sound pulses with no loss of energy. Should there be fractures, the sound paths are disturbed and energy is lost as the sound waves encounter the fractures. This loss of energy is shown on the log as decreasing amplitude. The amount of decrease, both in percent of total deflection and number of feet of interval, is indicative of the intensity of fracturing. (See Figure 56.) Generally a Gamma Ray Log is run in conjunction with the Amplitude Log.

§ **3.04(b) Cased Hole Application.** The Amplitude Log is used in cased holes to help evaluate the cementing operation. It is commonly called the "Cement Bond Log".

The sound pulse will travel through free hanging casing at full energy level and the log will record maximum amplitude. When cement surrounds the casing and is bonded to it, the energy level is decreased because the pipe is no longer free to vibrate. The log records reduced amplitude, with the amount of reduction depending on the quality of the bonding. There

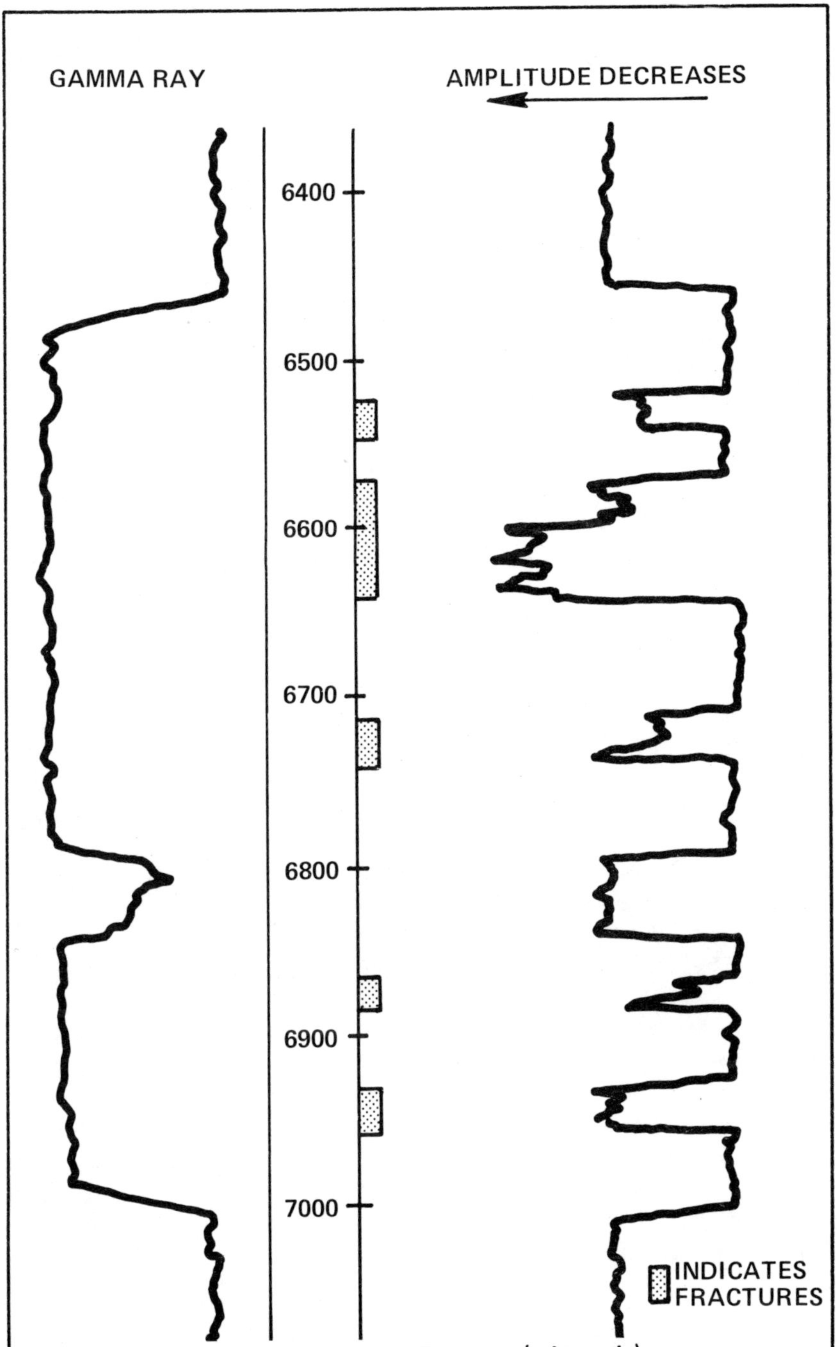

Amplitude Log In Fractures (schematic)

Figure 56

may be voids, gaps, and channels that can affect the amplitude measurements. The Gamma Ray Log and Casing Collar Log is usually run with this log as shown in Figure 57.

Sometimes in high velocity formations the Amplitude Log interpretation for cement bond quality can be confused by the signals coming through the rocks as well as the casing. A solution to this problem is the Variable Density Log or Micro-Seismogram. Figure 58 illustrates the principle of this log. 58-A is a sine wave as transmitted by the equipment. 58-B is the same sine wave with the negative portion removed. 58-C is the remaining positive half of the wave turned 90° so you are looking at the top of the peaks. The white gaps are where the negative peaks have been removed, and the black dashes are the full amplitude positive peaks. These are recorded on photographic film syncronized so that it will have the same depth scale as the other logs. The instrumentation is such that there are gradations in color from black at full amplitude to lighter grays as the amplitude decreases, hence the name "Variable Density Log". An example of this log is shown in Figure 59.

This log permits a study of amplitude variations and variations in time of arrival of the sound signals. This is helpful in cement bond determination, not only to the casing but to the formation. In uncased holes it is a valuable aid to fracture evaluation.

§ 3.05 The Thermal Decay Time Log or Neutron Lifetime Log

These logs differ from previously discussed Neutron Logs in that they use pulsed neutron sources. The pulsed sources permit the measurement of the neutron capture cross-section of the various formations, as well as the thermal decay time of the neutron density. Thermal decay time is related to the capture cross-section of formations, so the ultimate primary measurement in both cases is capture cross-section, or "Sigma" in logging terms. This measurement is very similar in appearence to open hole resistivity logs. It is, in fact, affected pretty much by the same things that affect resistivity measurements; that is, porosity, water salinity, water

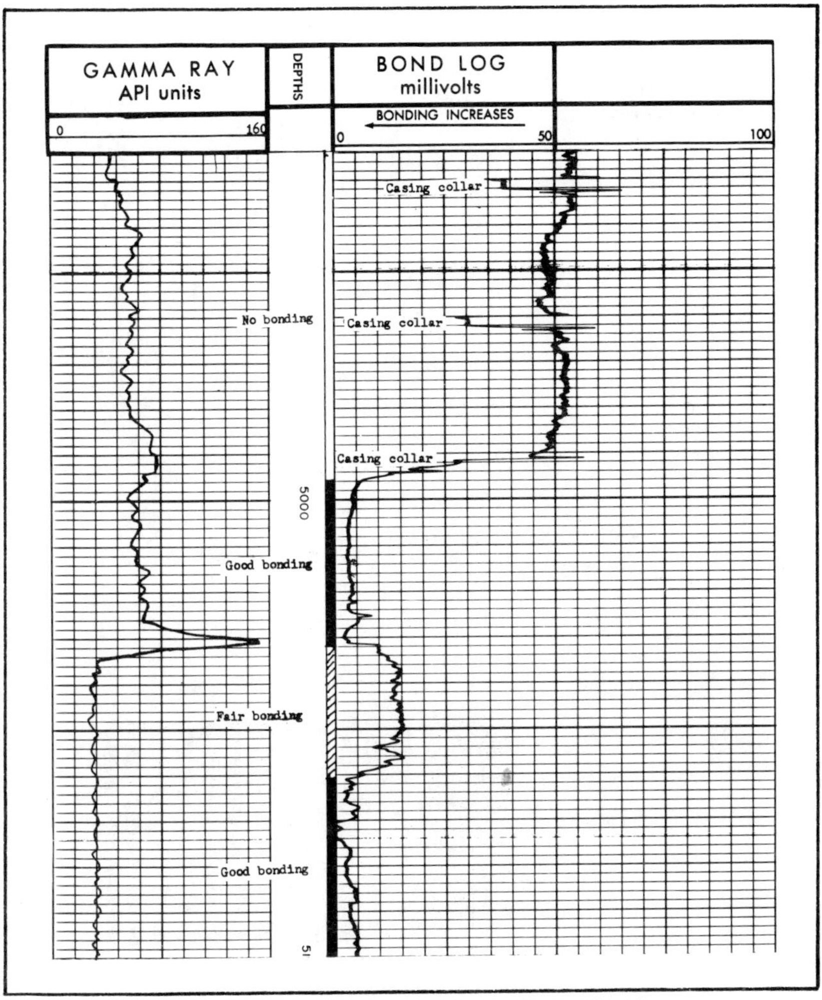

Amplitude Log For Cement Bond Determination
Figure 57

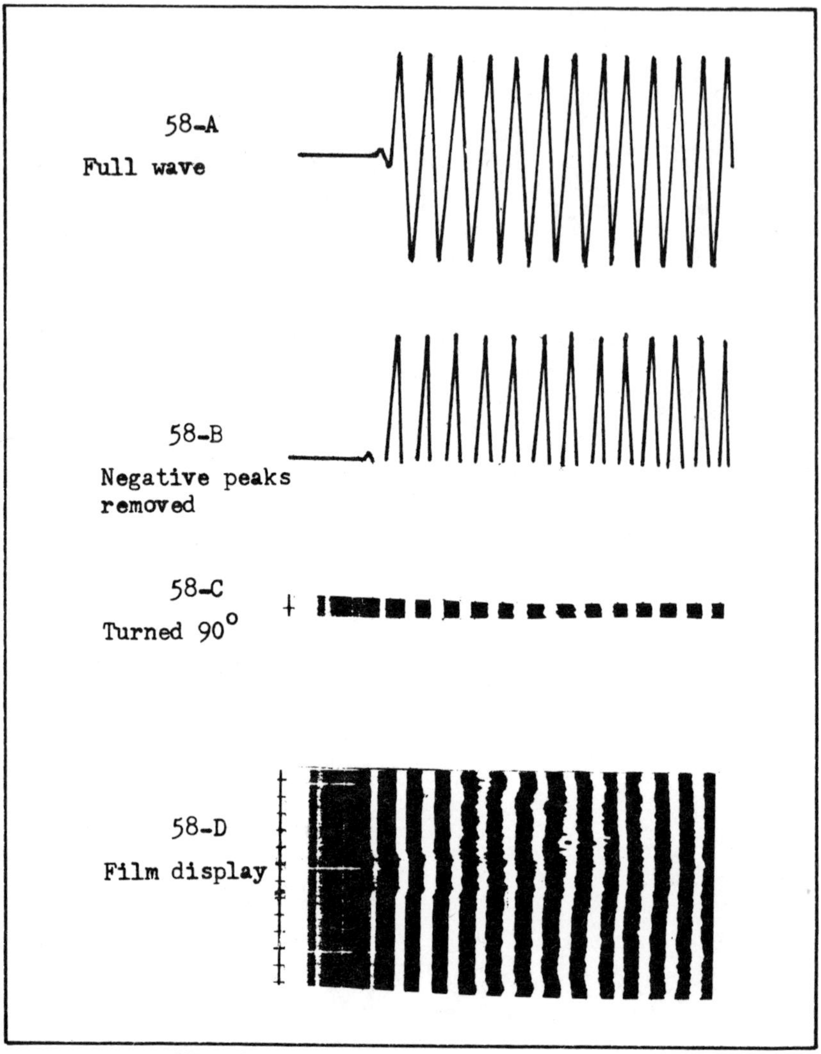

Micro-Seismogram or Variable Density Log

Figure 58

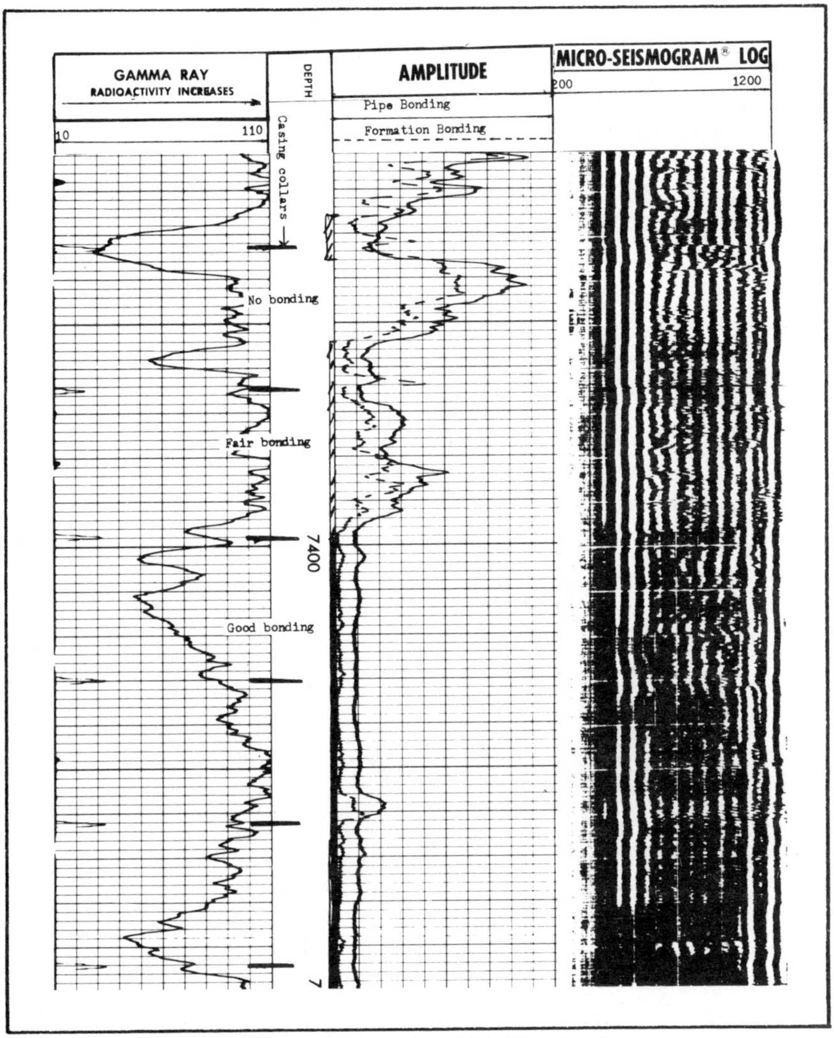

Micro-Seismogram or Variable Density Log
Figure 59

saturation, and shaliness.

Although this log may be run in uncased holes, its predominate use is in cased holes.

It is useful to evaluate old wells that may or may not have been logged previously, especially when looking for up-hole pay zones prior to abandoning the well, and also to monitor reservoir performance in producing wells by detecting changes in gas/oil/water contacts, or encroaching water tables. Production problems may be diagnosed through the use of these logs.

Quantitative analysis can be performed with a good degree of accuracy provided porosity, water salinity, shaliness, lithology, and hydrocarbon types are known or can be determined. This information is quite often available from other sources such as open hole logs, core data, and production history, to name a few.

§ 3.06 Production Logs

Production logs involve the use of a family of instruments that make measurements pertaining to the performance of a well after it has been completed.

The equipment that makes these measurements includes the following: Flow-meter (either continuous or packer type), high resolution thermometer, gradiomanometer, manometer, through tubing caliper, casing collar locator, fluid sampler, and radioactive tracer. The Gamma Ray may be run with these logs.

Basically, production logs answer the following questions: In a producing well how much of what is coming from where? In an injection well how much is going where?

In producing wells this information becomes important in two broad categories, reservoir performance and problems with the well. When a well has been completed and is performing as expected, everything is good. Very often production logs are run in such a well to provide a basis for future reference. The well may be logged periodically during its lifetime to monitor depletion, water encroachment, or any other change in production characteristics, which will provide a base from which to work should problems develop. There are many kinds of problems, such as casing leaks, tubing leaks, packer failures, communication behind pipe, plugged perforations, and failure of other mechanical completion hardware. By knowing what has occurred and where it has occurred makes the remedial work greatly simplified.

Production Logs also allow an evaluation of the effectiveness of stimulation of the formations, whether acidizing or fracturing.

In injection wells it is important to know which interval is taking fluid and the rate of injection. Production Logs provide this information and show the way for proper remedial work in case the injection profile is not as it should be.

Most of these logs are well understood by log analysts; however, two generally have specialists associated with them. The service companies usually have specially trained engineers to help the operators with the use of the Dipmeter and Production Logs, including the Thermal Decay Time and Neutron Lifetime Logs. There is much interpretive skill involved and the expertise of these engineers is very helpful.

Chapter Four

INTERPRETATION AND QUANTITATIVE ANALYSIS

We come now to the most important part of the whole logging procedure: Reading the story the logs have to tell.

When a set of logs has been run in a well we have at our disposal a great quantity of milli-volts, ohm-meters, micro-seconds, grams per cubic centimeter, and porosity units, depending on the combination of logs used. So far there is nothing that measures sandstone, limestone, dolomite, shale, oil, gas, or water, as such. All of these things affect the measurements made by the logging equipment, and it is the job of the log analyst to convert the one to the other.

The service companies that run logs make every effort to assure that the measurements presented to the well operator are as accurate as possible. The equipment is very reliable, and adjustments are made each time it is used in a well. Modern logs include calibration records which are made before and after logging, so that the analyst may check to be sure nothing happened during the operation to affect the measurements.

Properly run logs are probably the most impartial measurements made in a bore hole. They do not "lie", if you will, since the instruments are merely making the measurements they are designed to make in the environment presented to them. It is up to the people using them to read their story.

In the early years of logging, and especially prior to focused type logs, reading the logs was an interpretative procedure and involved a great deal of "art". With the advent of focused logs, measurements became very accurate and mathematical techniques were developed so that analysis of logs has now become pretty well a "science" in its own right. It is not infallible. However, most of the errors made now are due to the interpretation of the results of the analysis, and not the analysis itself. More about this later.

Let us start our discussion on interpretation with the old style Electrical Log. This log was just about it until the late 1940's, when focused logs began to arrive.

§ 4.01 Electrical Logs

The Electrical Log consists of a spontaneous potential curve and two or more resistivity curves. The resistivity curves were of different electrode spacings and arrangements. The purpose of this was to have curves that have different depths of investigation. A shallow investigation curve measures close to the bore hole and is influenced by the invaded zone of permeable formations. A deep investigation curve measures far from the bore hole and is influenced by the non-invaded zone, hopefully to measure true resistivity (Rt). An intermediate investigation curve measures somewhere between the shallow and deep curves; this curve helps to determine how deep the formation is invaded and whether or not the deep curve is predominately true resistivity (Rt).

In those days quantitative analysis, as such, had not arrived yet. The interpretation of the log was more a case of deciding whether a zone should make oil or gas or water. This usually was accomplished by comparing the appearance of zones on the log to what you knew the zone looked like on a log where the zone was water filled. This is where experience and "art" entered the picture.

If you knew, for example, that a particular formation looked like Figure 60 where it was water filled, and you had a log nearby that you were making an interpretation on where the same formation looked like Figure 61, you would draw the conclusion that the formation should produce oil or gas.

This conclusion will be based on an assumption that may or may not be

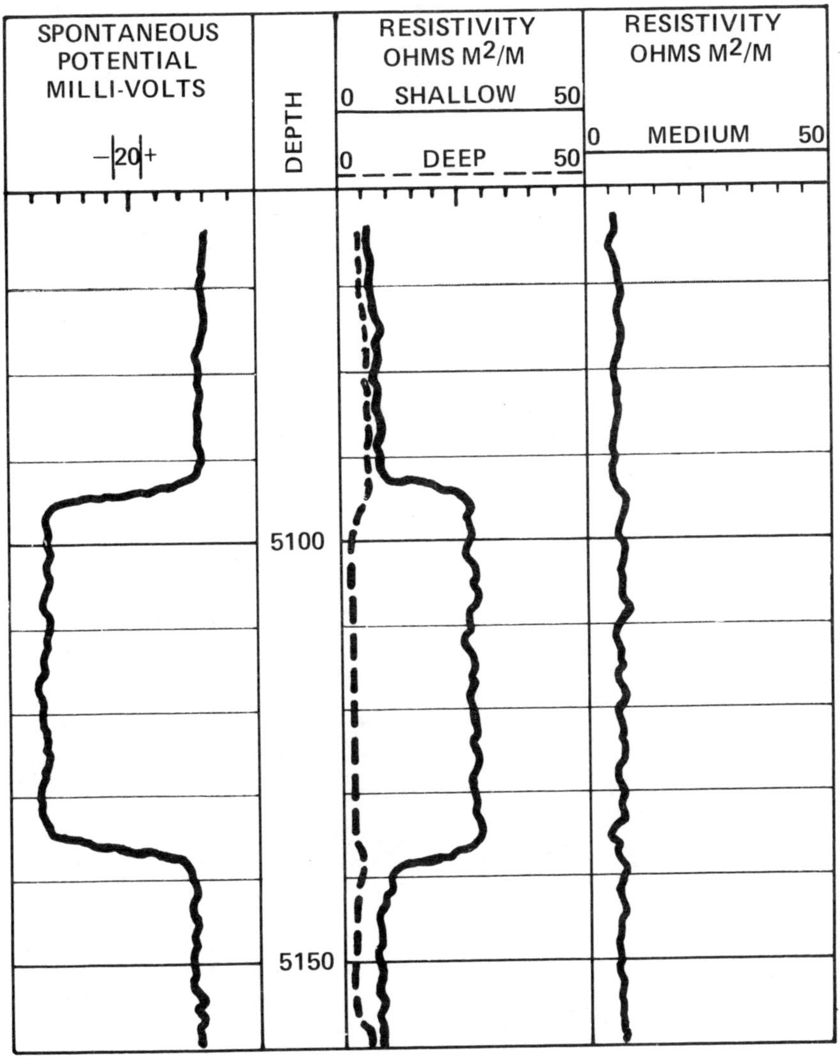

Water Filled Formation (schematic)
Figure 60

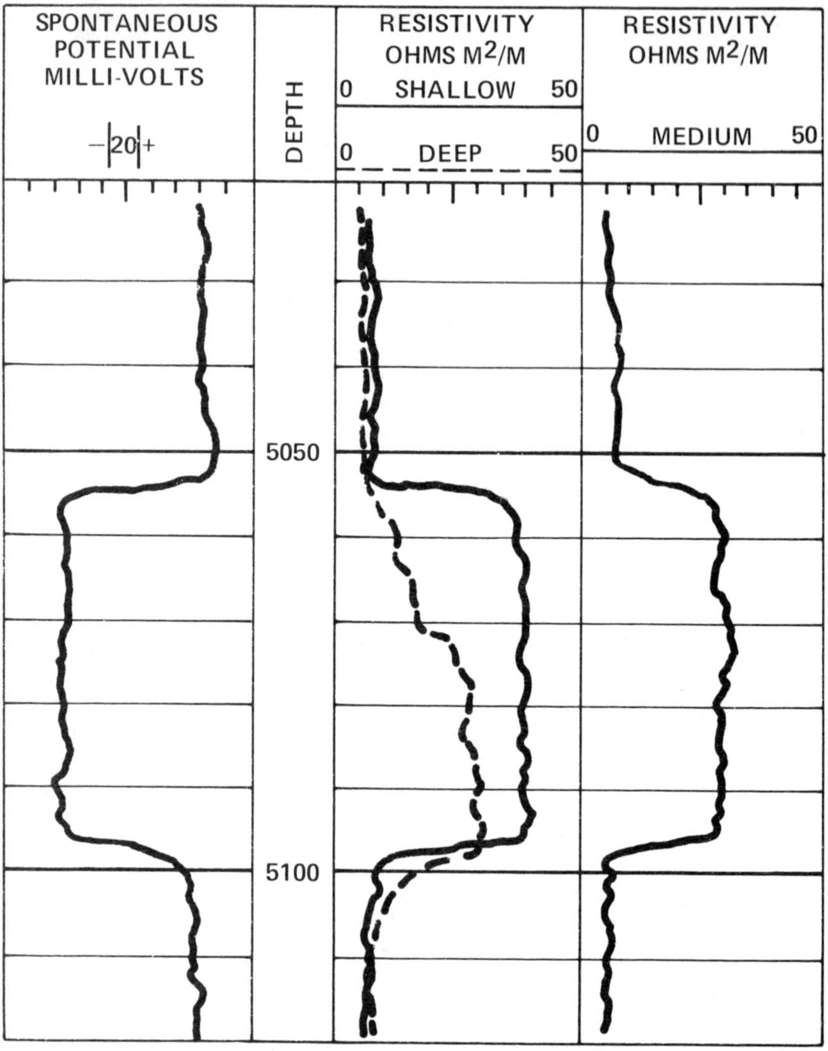

Oil Or Gas Filled Formation (schematic)
Figure 61

valid, which is that the porosity in both instances is the same or very similar. Should the formation shown in Figure 61 have somewhat lower porosity, the resistivity could be due to higher formation resistivity factor (F) and not due to oil or gas. Consequently, the interpretation that the formation in Figure 61 is oil or gas would be erroneous.

This method of interpretation, commonly called the "Eyeball" method, dominated the use of the Electrical Log for a long time.

In fact, it was common practice for the logging engineers to carry a loose leaf binder filled with examples of formations both water filled and oil or gas productive to compare to logs being interpreted. The books were constantly being added to as more pay zones were found and experience was gained.

Very early on many people realized the importance of electrical logs and much effort went into learning how they could be used more effectively. The search was for methods to analyze the logs quantitatively; that is, to arrive at reasonably accurate values for porosity and water saturation of the formations. One of these early investigators was Mr. M. P. Tixier with Schlumberger, who developed one of the most successful techniques. This was reported in a paper entitled "Electric Log Analysis in The Rocky Mountains", which was written in the 1940's. The method proved to be reliable and was very widely used.

Use of this method required knowledge of Ri, Rt, Rmf, and Rw. Ri is determined from the shallow investigation curve, and Rt from the deep investigation curve. Rm (mud resistivity) is measured at the time the log is run and recorded on the log heading, and from this Rmf is determined. Rw is from catalogs of measured values or determined from the S. P. curve. Two charts are used to solve the mathematics, and are shown in Figures 62 and 63.

To demonstrate the use of this interpretation technique let's look at the idealized log in Figure 64. It is an Electrical Log consisting of a Spontaneous Potential curve, a 16″ Normal curve (shallow investigation), a 64″ Normal curve (intermediate investigation), and a 19′ Lateral curve (deep investiga-

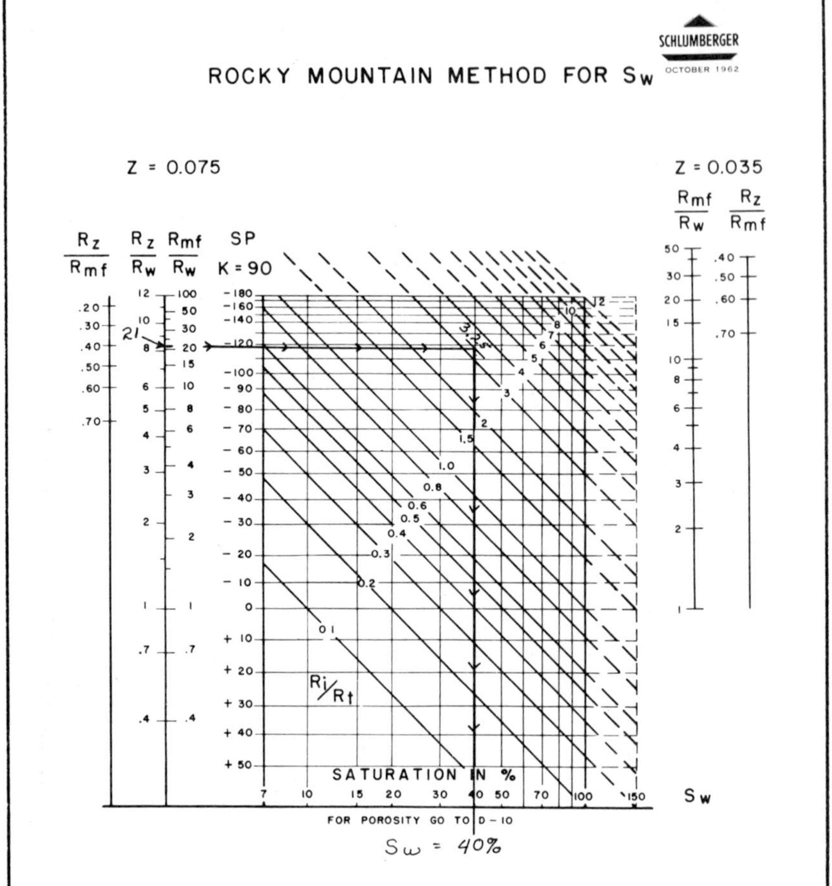

REFERENCE: Tixier, M. P. "Electric Log Analysis in the Rocky Mountains." *Oil & Gas Journal*, June 23, 1949

Figure 62

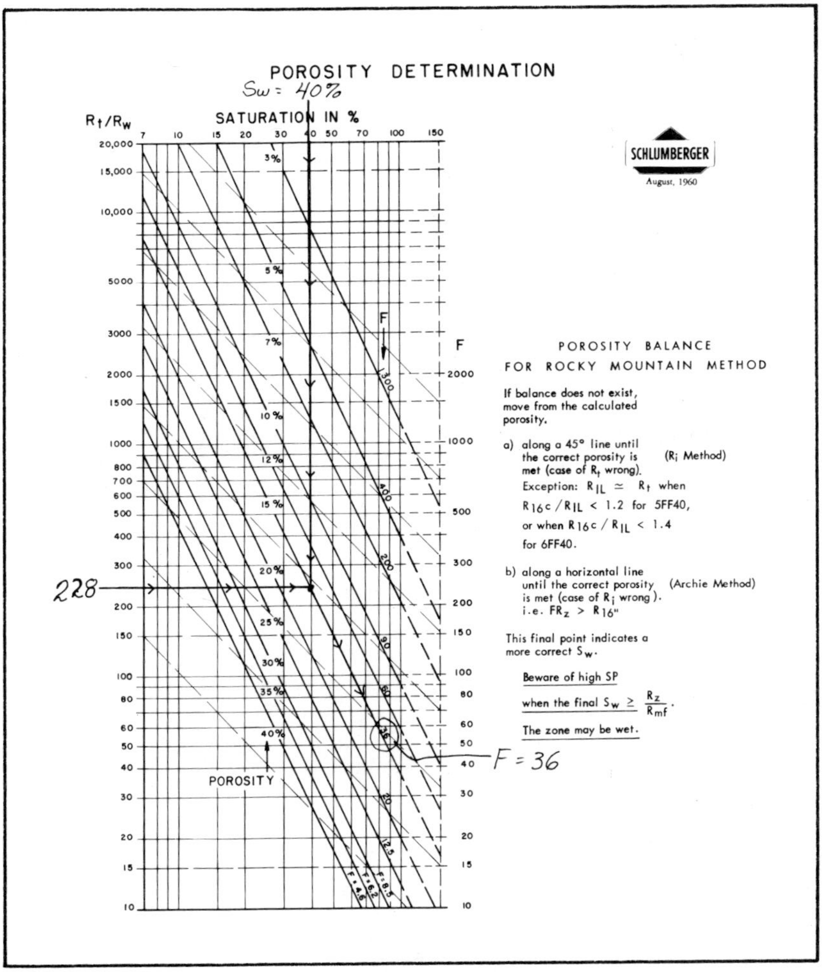
PORSITY DETERMINATION
Sw = 40%
SATURATION IN %
R_t/R_w
7 10 15 20 30 40 50 70 100 150
20,000
15,000
10,000
5000
3000
2000
1500
1000
800
700
600
500
400
300
200
150
100
80
60
50
40
30
20
15
10
3%
5%
7%
10%
12%
15%
20%
25%
30%
35%
40%
POROSITY
228
F
1300
400
200
90
60
36
20
12.5
F=4.6
F=6.2
F=8.5
F
2000
1000
500
300
200
150
100
80
60
50
40
30
20
15
10
F = 36
SCHLUMBERGER
August, 1960
POROSITY BALANCE
FOR ROCKY MOUNTAIN METHOD
If balance does not exist, move from the calculated porosity.
a) along a 45° line until the correct porosity is met (case of R_t wrong). (R$_i$ Method)
Exception: $R_{IL} \simeq R_t$ when
$R_{16c}/R_{IL} < 1.2$ for 5FF40,
or when $R_{16c}/R_{IL} < 1.4$
for 6FF40.
b) along a horizontal line until the correct porosity is met (case of R_i wrong). (Archie Method)
i.e. $FR_z > R_{16"}$
This final point indicates a more correct S_w.
Beware of high SP
when the final $S_w \geq \frac{R_z}{R_{mf}}$.
The zone may be wet.

Figure 63

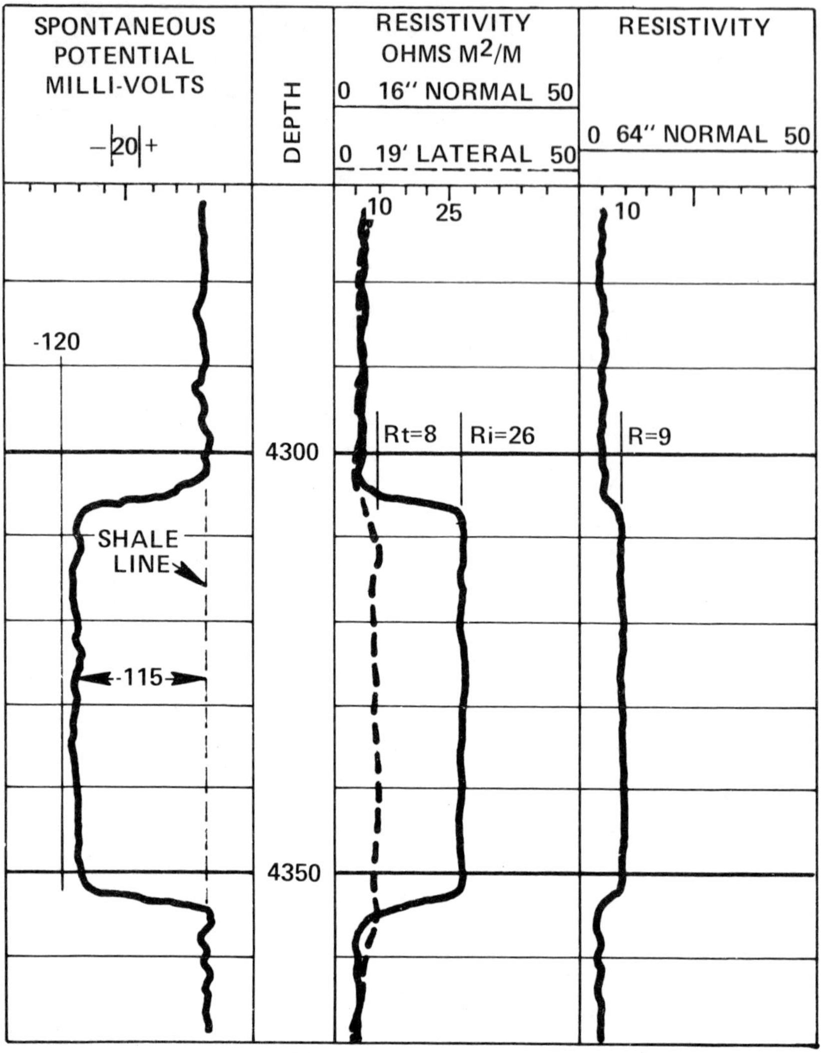

Idealized Electrical Log
Figure 64

tion). The 16″ Normal indicates Ri = 26 ohm-meters, the 19′ Lateral indicates R_t = 8 ohm-meters. Please note that the 64″ Normal = 9 ohm-meters which is fairly close to Rt, indicates that invasion is probably not very deep. Rmf has been determined to be .75 ohm-meters at 125°, and Rw is .035 ohm-meters at 125°.

To enter the chart in Figure 62 we need two ratios, Rmf/Rw and Ri/Rt. From our known data these are: .75/.035 = 21, and 26/8 = 3.25. Enter the chart on the left vertical scale at 21 and go horizontally right until just past the diagonal line 3 to 3.25. At this point drop straight down to the bottom scale for saturation, and find 40%. Now to find porosity we need to enter the chart in Figure 63. One other ratio is needed and we have the necessary numbers to make it. They are Rt and Rw. Rt = 8 and Rw = .035 at 125° so the ratio Rt/Rw = 8/.035 = 228. Enter the chart on the left vertical scale Rt/Rw = 228 and go right horizontally. Enter the top horizontal scale for saturation at 40%. Drop down to the intercept with the horizontal line from Rt/Rw = 228. They intercept at a diagonal line equal to 15% porosity. This interval then has porosity = 15% and water saturation = 40% and should produce oil or gas. Please notice that if you follow the diagonal line for 15% porosity downward, you will come to the number 36. This is the formation factor for 15% porosity. This method provides for another solution for porosity so that it is possible to check on whether or not the interpretation of the log was reasonable. It is known that $Ri = \frac{F \times Rz}{Si^2}$ Where Rz = the resistivity of the fluid in the invaded zone, Si = saturation of the fluid in the invaded zone, and F, of course, is the formation resistivity factor. Rz is determined from the chart in Figure 62 using the scale next to Rmf/Rw. In our example Rmf/Rw = 21 so Rz/Rw = 8.3. Rz = Rz/Rw = 8.3 x .035 = .29. Mr. Tixier determined through his research that $Sw = Si^2$. We now have all the necessary data to make our check. If we rearrange the formula $\mathbf{Ri} = \frac{\mathbf{F \times Rz}}{\mathbf{Si^2}}$ to $\mathbf{F} = \frac{\mathbf{Ri}}{\mathbf{Rz}} \mathbf{\times Si^2}$ and use our data we will find $F = \frac{\mathbf{26}}{\mathbf{.29}}$ **x .40** which equals 35.8. This is the same as the formation factor for 15%

porosity. Therefore, our interpretation must be reasonably correct.

Needless to say, all interpretation is not as easily accomplished as this example. So many factors affected the Electrical Log curves that it was very difficult to determine accurate values necessary to use the method. There were effects from the bore hole, bed thickness, bed resistivity, mud resistivity, and invasion. In an attempt to overcome these problems "Resistivity Departure Curves" were calculated. There were several books of these curves presented to the industry, which were complicated and somewhat difficult to use, and very often a "fit" could not be obtained to accommodate the conditions for the log you were working on.

§ 4.02 Focused Logs

Focused logs were invented and developed and pretty well solved many of these problems. We will discuss each one as we progress through this chapter.

From Mr. Tixier's interpretation method we can see that it would be very helpful if there were a way to determine porosity of the formations in place. One of the first approaches to making this determination was the Micrologᵀ. The mechanics of the Microlog was covered in the porosity chapter, and we will now cover the analysis. An idealized Microlog is shown in Figure 65.

As mentioned previously the factors that affect the Microlog are: Mud filtrate resistivity (Rmf), Mud cake resistivity (Rmc), Mud cake thickness (hmc), saturation of residual hydrocarbons (Shr), and formation resistivity factor (F). Rmf and Rmc may be measured from a mud sample and assumed to be correct at various depths in the well. They may also be estimated from the mud resistivity if no other information is available. The mud cake thickness may be determined from the caliper curve. Saturation of residual hydrocarbons must be estimated. The purpose of the solution of the Microlog is to obtain formation resistivity factor (F). The mathematics to solve the Microlog are complex, so two charts were devised that simplify the procedure. These charts are shown in Figures 66 and 67.

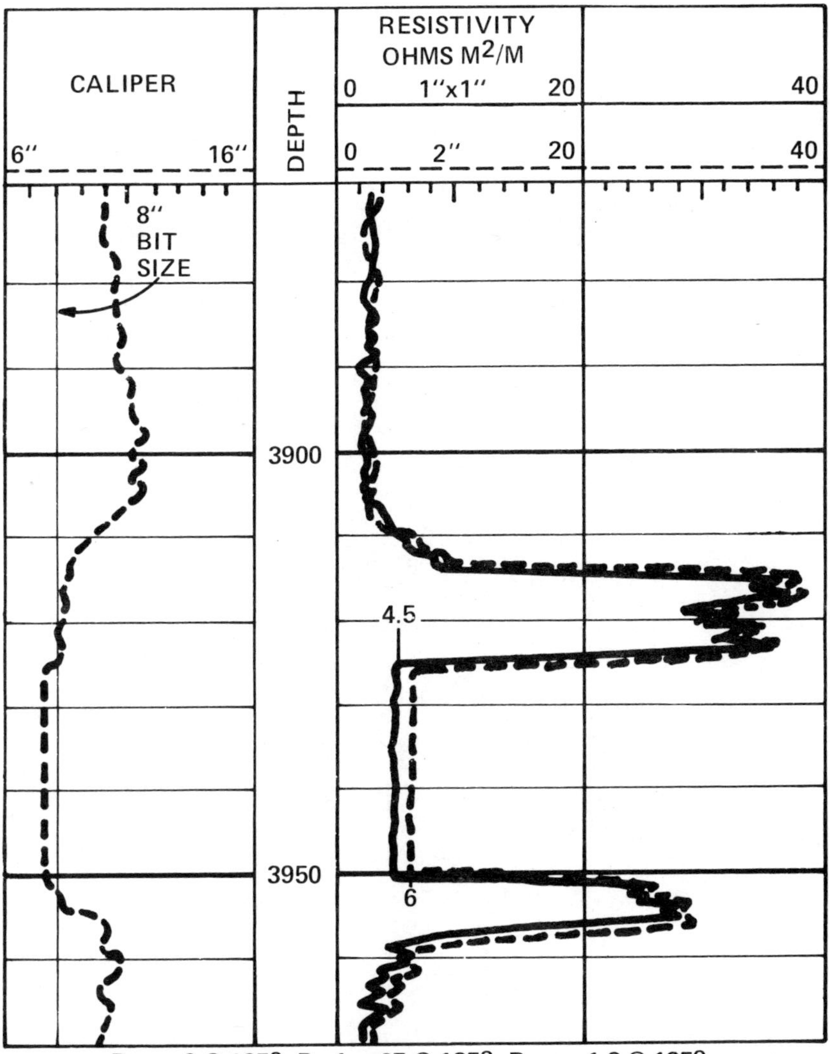

Rm = .9 @ 125° Rmf = .67 @ 125° Rmc = 1.2 @ 125°

Idealized MicroLog
Figure 65

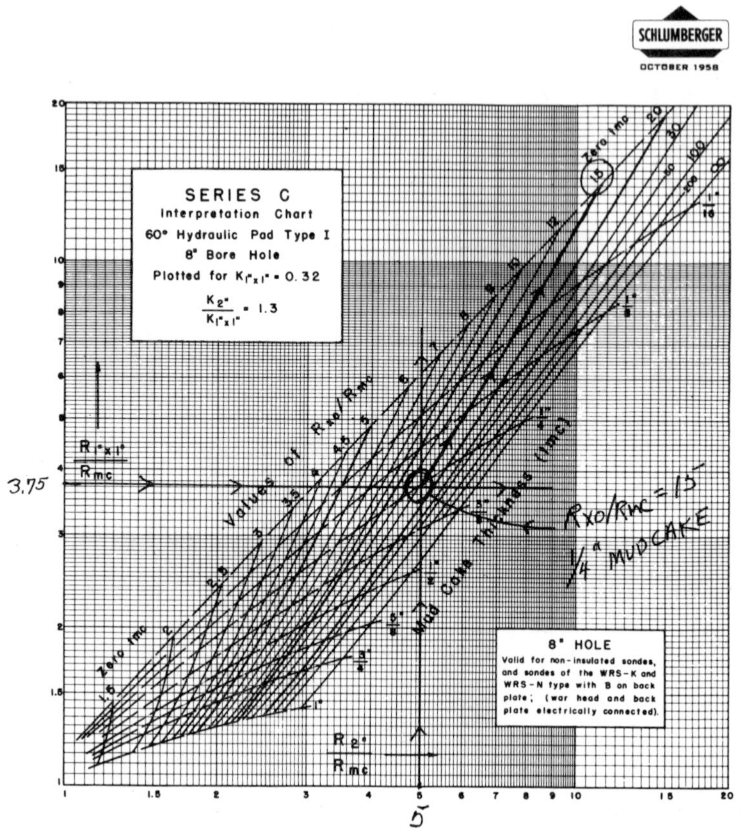

NOTE: An incorrect R_{mc} will displace the points in the chart along a 45° line passing through that point. In certain cases this can be recognized when the mud cake thickness is different from direct measurement by the microcaliper. To correct, move the point at 45° to intersect the known t_{mc}. Read $R_{xo}/R_{mc'}$ at the middle and $R_{2''}/R_{mc'}$ at the bottom of the chart. Divide the log value of $R_{2''}$ by the ratio $R_{2''}/R_{mc'}$ to find $R_{mc'}$. Now, $R_{xo}/R_{mc'}$ times $R_{mc'}$ equals R_{xo}.

Figure 66

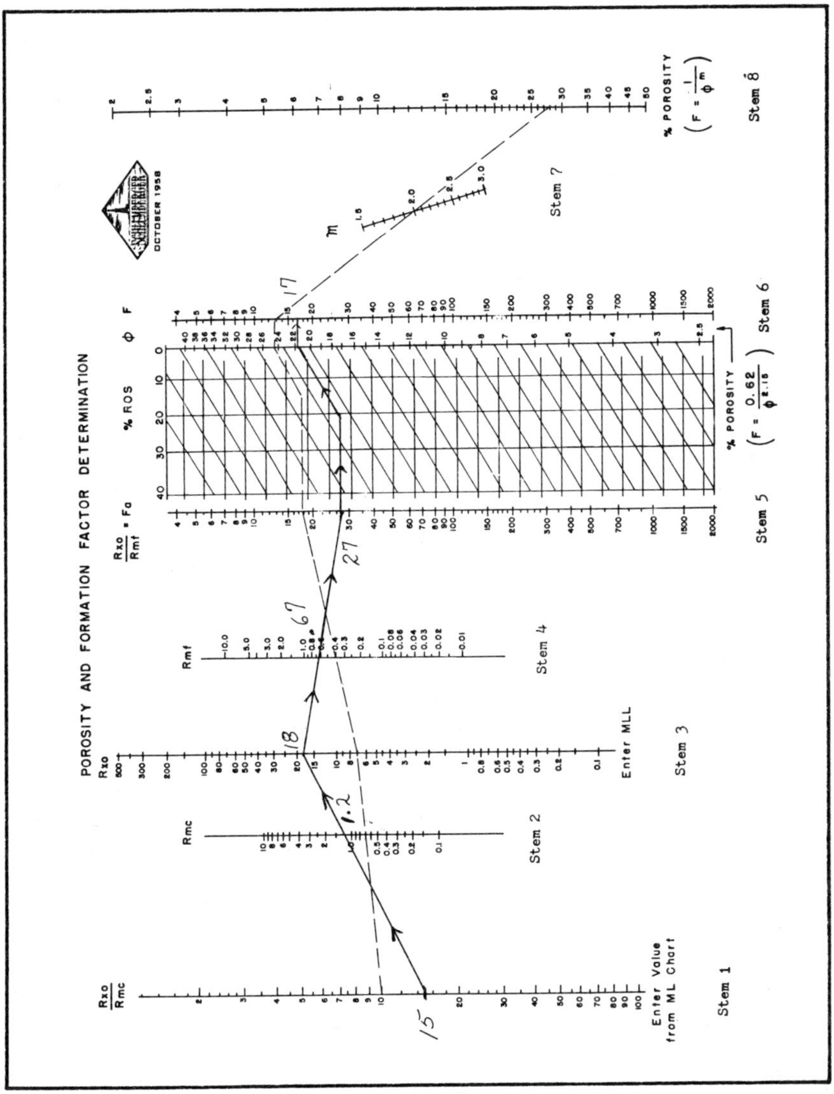
POROSITY AND FORMATION FACTOR DETERMINATION
OCTOBER 1958
Rxo/Rmc
Enter Value from ML Chart
Stem 1
Rmc
Stem 2
Rxo
Enter MLL
Stem 3
Rmf
Stem 4
Rxo/Rmf = Fa
Stem 5
% ROS
% POROSITY
Stem 6
m
Stem 7
% POROSITY
Stem 8

Figure 67

The data necessary to use the Microlog charts are: Resistivity of the 1″ x 1″ micro-inverse, resistivity of the 2″ micro normal, Rmc, Rmf, and Shr. From the Microlog in Figure 65 we find the resistivity of the 1″ x 1″ = 4.5, the resistivity of the 2″ = 6, Rmc = 1.2 @ 125°, Rmf = .67 @ 125°. Shr we will estimate later.

To enter the chart in Figure 66 requires two ratios, R1″x1″/Rmc and R2″/Rmc. From our data we have 4.5/1.2 = 3.75 and 6/1.2 = 5. Enter the chart from the left scale at 3.75 and extend a line to the right, enter the bottom scale at 5 and extend a line up till it intercepts the other line. The intercept will be a point equal to a value for Rxo/Rmc and a value for mud cake thickness. Our point will fall at Rxo/Rmc = 15 and mud cake thickness = 1/4″. The mud cake thickness is similar to that indicated by the caliper. That is all we get from this chart, so now we are ready to enter the chart in Figure 67. Enter this chart at the left side on stem #1 Rxo/Rmc = 15. Go through Rmc = 1.2 on stem #2 to stem #3 and find Rxo = 18. This is the resistivity of the flushed zone. Go through Rmf = .67 on stem #4 to stem #5 and find Rxo/Rmf = 27. This is F apparent, and would be F if the zone were water filled. If the zone has oil or gas in it then there will be some residual hydrocarbon saturation. This chart is old (1958), and at that time this was called ROS (residual oil saturation). If we assume ROS = 20 % (which is not an unusual figure), we will proceed across to the line for 20 %, then join the family of diagonal lines upward to the right and find porosity is about 21 to 22 %. At this porosity F = 17 if we use the Humble relationship shown at the bottom of the graph ($F = \frac{.62}{\phi^{2.15}}$). This is good for most sandstones. Should one wish to vary the cementation factor "m" this chart permits that by going through the selected "m" on stem 7 to stem 8.

From this procedure we can see that the Microlog solution is complicated and subject to errors from several sources and assumptions. This is why it is seldom used as a porosity log anymore. Its use is pretty well localized for this purpose; however, it is often used in areas as a net footage of pay measurement.

Let's put the older style logs to rest now and proceed to the modern logs -- the focused resistivity measurements and the porosity sensitive logs.

§ 4.03 Focused Resistivity Measurements

The focused resistivity devices, such as Laterolog (Guard Log) and Induction Logs, have practically eliminated the effects of the bore hole and formation thickness (greater than about 6 feet), and permit correction for invasion effects in most cases. However, they still measure resistivity with no way to tell if the resistivity is due to oil or gas or tight rock. Therefore, measurement of porosity is of prime importance.

Because of this, and because of the fact that nearly all analytical techniques involve the use of porosity and the response of the various logs, they will be discussed first.

§ 4.03(a) The Sonic Log. The Sonic Log (Acoustic Log), as discussed in the chapter on porosity, is solved for porosity by using Mr. Wyllie's time average formula, which is:

$$\text{Porosity} = \frac{\Delta t_{log} - \Delta t_{matrix}}{\Delta t_{fluid} - \Delta t_{matrix}}$$

Rather than go through this for each interval of interest a chart is available to simplify the procedure. (See Figure 68.) With this chart any transit time in micro-seconds can be converted to porosity by drawing lines. Let's solve the Sonic Log in Figure 69 to demonstrate the technique. This is the log through a limestone section broken by shales a 6226′ to 30′; 6248′ to 52′; 6291′ to 6311′; and 6344′ to 50′. The predominate interval of porosity is indicated a 6260′ to 73′. The log is scaled 40 micro-seconds at the right side of the log, 70 micro-seconds at the center, and 100 micro-seconds at the depth column. There are 3 micro-seconds for each vertical division of the log, so a pretty good average value for our interval of interest is 67 micro-seconds. We could work the time average equation to get porosity, but we will use the chart in Figure 68.

Find 67 on the bottom scale (Figure 68) and strike a line upward through

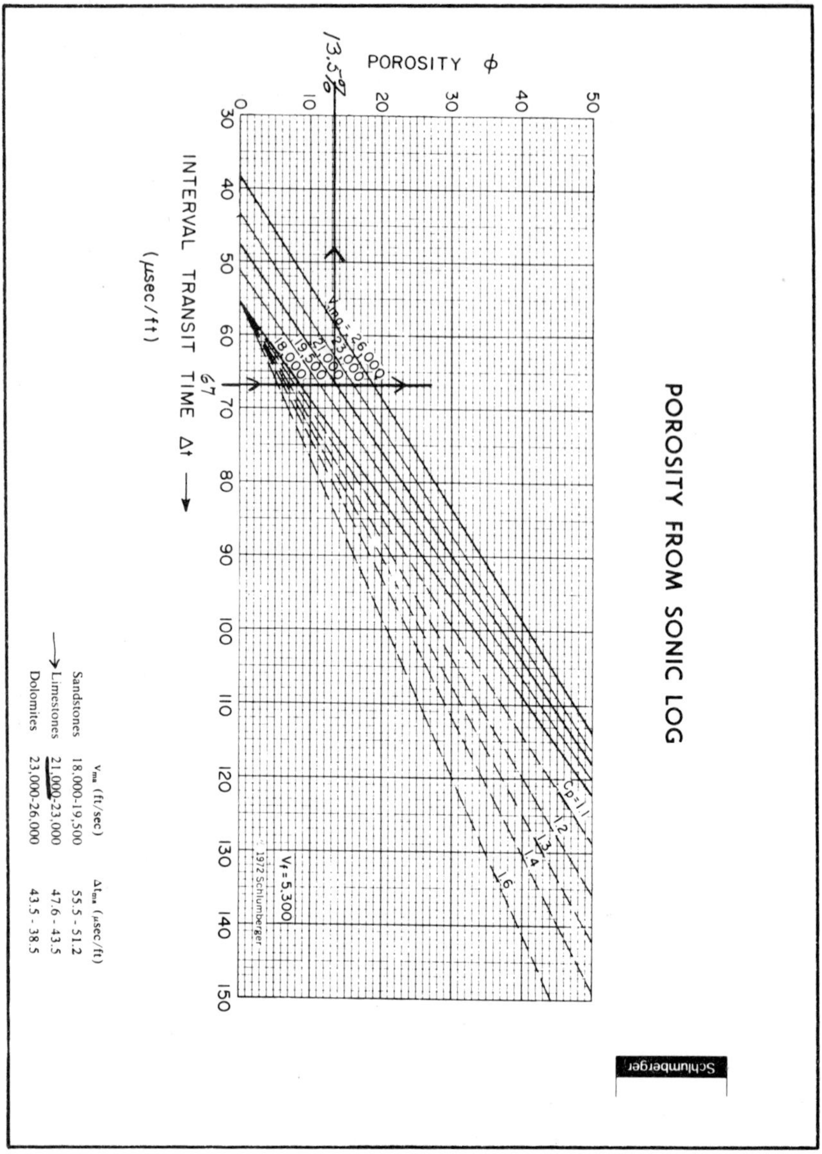

Chart To Solve Sonic Log
Figure 68

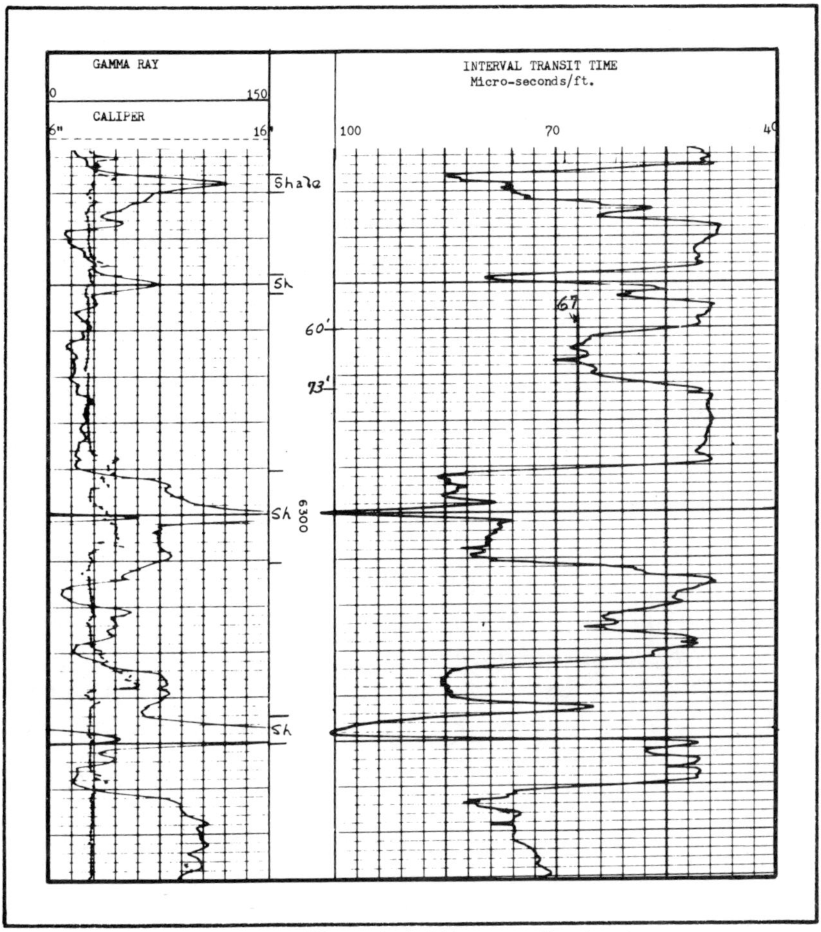

Sonic Log Through A Limestone Section
Figure 69

the family of curves. Since this is limestone, and most limestones have a matrix velocity of 21,000′ per second, we will solve at this value. Where the upward line intercepts the 21,000 diagonal line strike a line left to the porosity scale and find about 13.5 %.

Had this interval been sandstone we would have stopped at the 18,000′ diagonal and found porosity = 8.5 %. If it had been dolomite we would have gone to the 23,000′ diagonal and found porosity = 16 %. From this you can see the importance of knowing something about the lithology. However, this chart allows one to vary the rock type and see the effect on the porosity determination. In some geological provinces the rocks are much less compacted by overburden, and this makes necessary a compaction correction on the Sonic Log. That is what the dashed curves are at the right side of the 18,000′ diagonal. Local conditions will dictate their use, so anyone working in "soft rock" areas, the Gulf Coast, for example, should seek advice from the logging company engineers as to or about using this correction.

As mentioned previously, shaliness in the formation will usually cause the Sonic Log to indicate porosity higher than it should be. To obtain the correct porosity a correction must be made to eliminate the shale effect. There are many approaches to making this correction; some of them are complex and some even involve computers. There is a method that is relatively simple which is widely used and generally accepted. It involves the use of alpha (α), which is the SP reduction factor due to shaliness. If we observe the SP in clean water sands we will see the maximum value available for the particular Rmf/Rw ratio in that well. This is called the Static Spontaneous Potential (SSP). Then, if we observe the SP in the shaley sand of interest we will see the Pseudo Static Spontaneous Potential (PSP) which is the maximum SP in the shaley sand for the same Rmf/Rw ratio. Figure 70 illustrates this procedure with an idealized log. Alpha (α) is determined by the ratio PSP/SSP. So in our example PSP = -80 milli-volts and SSP = -100 milli-volts and alpha = .8.

Correction of the apparent porosity from the Sonic Log in a shaley sand

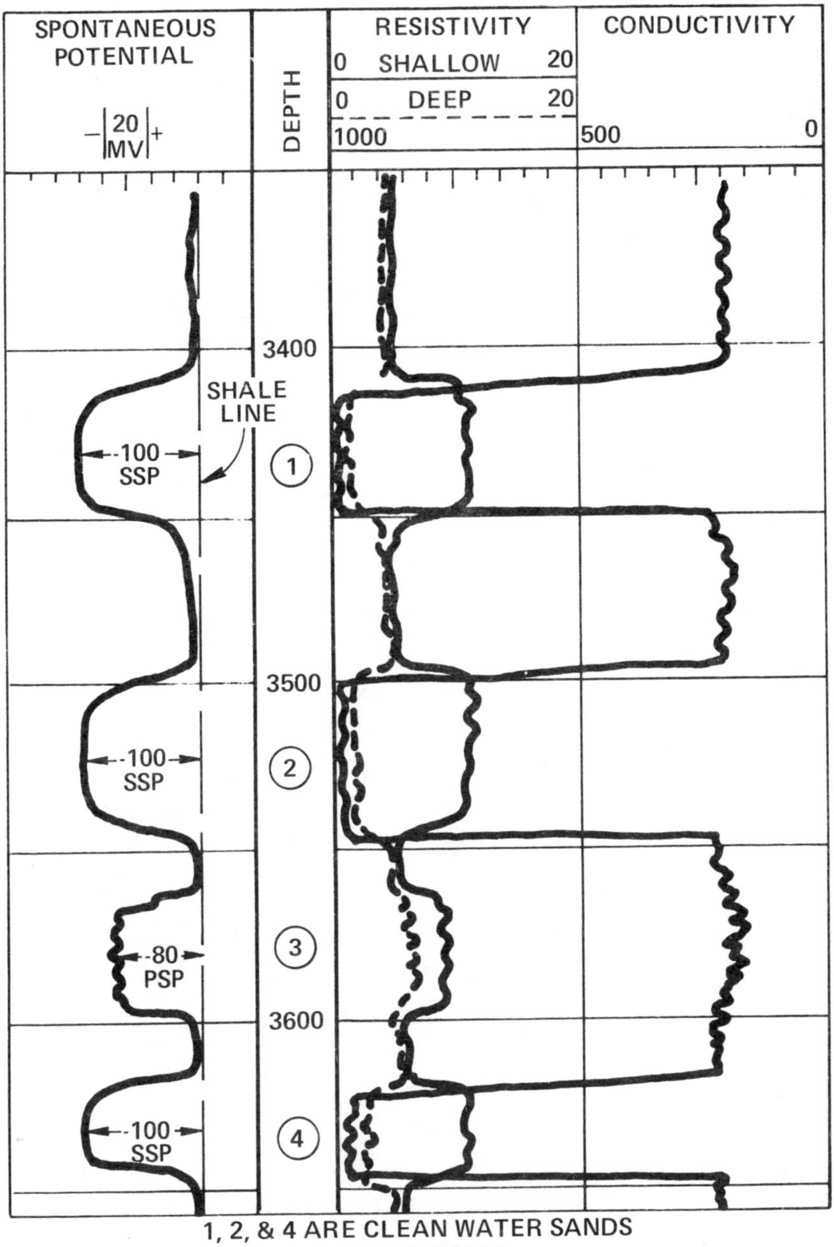

1, 2, & 4 ARE CLEAN WATER SANDS
3 IS SHALEY SAND

Static SP vs Pseudo Static SP (schematic)
Figure 70

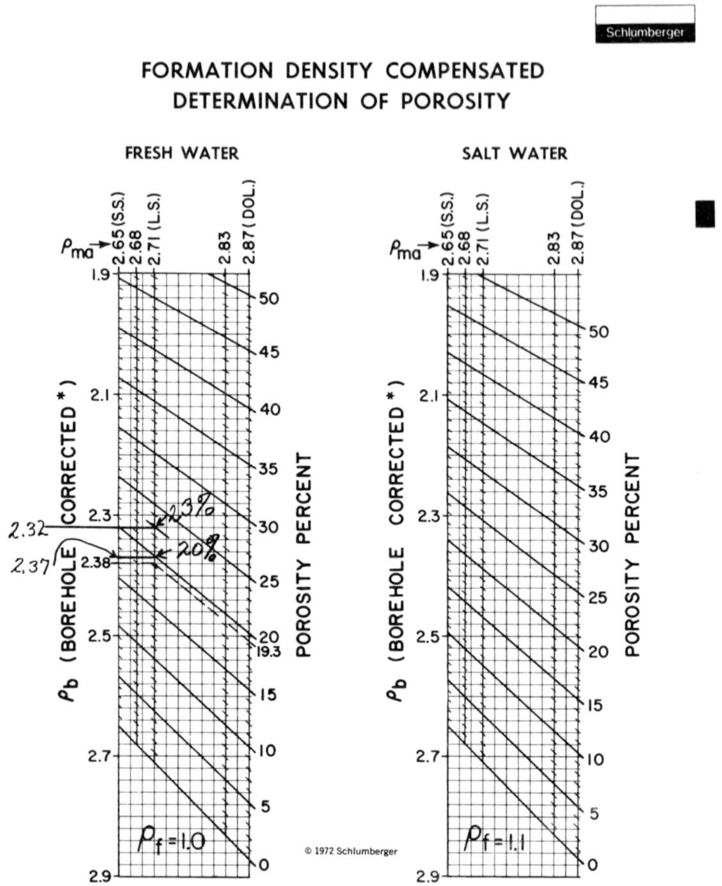

* See Por-5b, page 18.

Example: $\rho_b = 2.38$ in limestone lithology ($\rho_{ma} = 2.71$), fresh mud ($\rho_f = 1.0$).

Solution: $\phi_D = 19.3$.

Compensated Density Log Chart
Figure 71

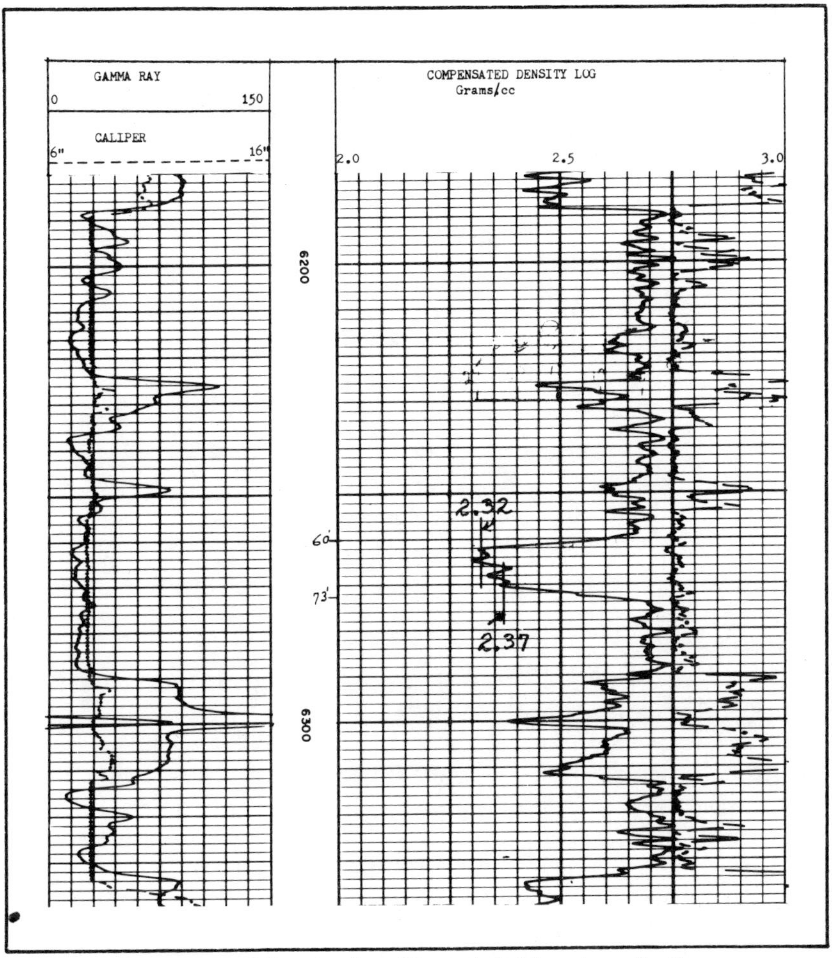

Compensated Density Log Through A Limestone Section
Figure 72

involves the use of this ratio: **Correct porosity** = $\frac{\textbf{Ø Sonic Log}}{\textbf{2 - alpha.}}$. If we find the Sonic Log solves to 23 % porosity on the chart in Figure 68 and alpha = .8 then correct porosity = .23/1.2 (2-.8) or .19, so our correct porosity is 19 %.

There is another condition in which the Sonic Log porosity may vary from the actual value. This is the case where carbonate rocks have secondary porosity. Detecting this condition involves the use of an additional porosity sensitive device as well as the Sonic Log. The Density Log is quite often useful in this situation.

§ 4.03(b) Density Log

First we will discuss the solving of the Density Log for porosity, and then we will illustrate the secondary porosity procedure. Porosity from the Density Log is equal to $\frac{\rho g - \rho b}{\rho g - \rho f}$. As with the Sonic Log, there is a chart that simplifies this solution. It is shown in Figure 71. The rock type determines the grain density to be used. Figure 72 is an example of a Compensated Density Log through the same limestone interval as the Sonic Log in Figure 69. The porosity interval is from 6260′ to 73′. Bulk density in this interval varies from 2.32 to 2.37.

The chart in Figure 71 has two sections. One is used with fresh water (ρ fluid = 1), while the other is used with salt water (ρ fluid = 1.1). Since the depth of investigation of the Density Log is very shallow, it will measure in the flushed zone where mud filtrate will be the predominate fluid. In our example we will use the fresh water section to solve for porosity. Enter the chart on the left scale at 2.32 and 2.37 and strike lines horizontally right where they intercept the line for ρ g=2.71 (limestone), and read porosity by interpolating between the diagonal porosity lines and find 20 to 23% porosity for the zone in our example.

In this same interval the Sonic Log indicated the porosity was 13.5%. The difference is due to secondary porosity. The secondary porosity index

(SPI) is equal to Density Log porosity (ØFDC) minus Sonic Log porosity (ØSL). In our example SPI = 6.5 to 9.5%.

If we had solved for porosity at sandstone grain density of 2.65 we would have porosity = 17 to 20%. If we had solved for porosity at dolomite grain density of 2.87, we would have porosity = 27 to 29.5%.

As you can see from the Sonic Log and Density Log discussions knowledge of the rock type is important in order to arrive at the proper porosity value. If we know the geological column in the area in which we are working, or if we are fortunate enough to work with formations of single mineral composition, we will not have too much trouble. However, if we are working with formations that vary in their composition such as mixtures of sandstone and shale, limestone and dolomite, or sandstone and limestone, and throw in the possibility that gas could be involved, much difficulty could be encountered in determining correct porosity values if we have only one porosity sensitive log to work with.

§ 4.03(c) Neutron Log

The Neutron Log, and specifically the Compensated Neutron Log, has purposely not been discussed individually because there is no need to solve it for porosity. The equipment does this automatically and scales the curve on the log in linear porosity values. The only requirement is that the rock type must be selected by the logging engineer. He may select sandstone, limestone, or dolomite, and change them at will as the log is being run, with proper notations as to where the changes were made. Again, the rock type becomes involved.

§ 4.03(d) Porosity Sensitive Device Combinations

One of the most widely-used methods to solve the mineral mixture and/or gas effect problem is to use a combination of porosity sensitive devices. The combination that is probably preferred the most is the Compensated Neutron Log and the Compensated Formation Density Log. When these two logs are run in combination they will both be solved and the

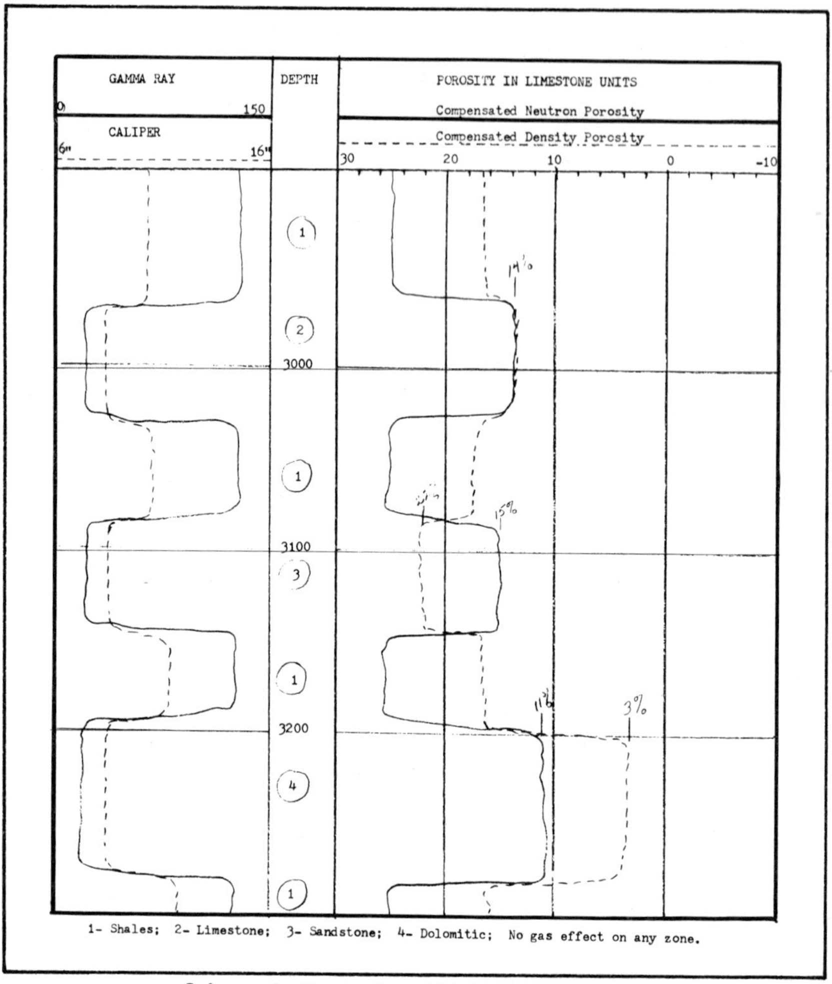

Schematic Illustration Of Lithologic Effect On Responses Of Neutron And Density Logs

Figure 73

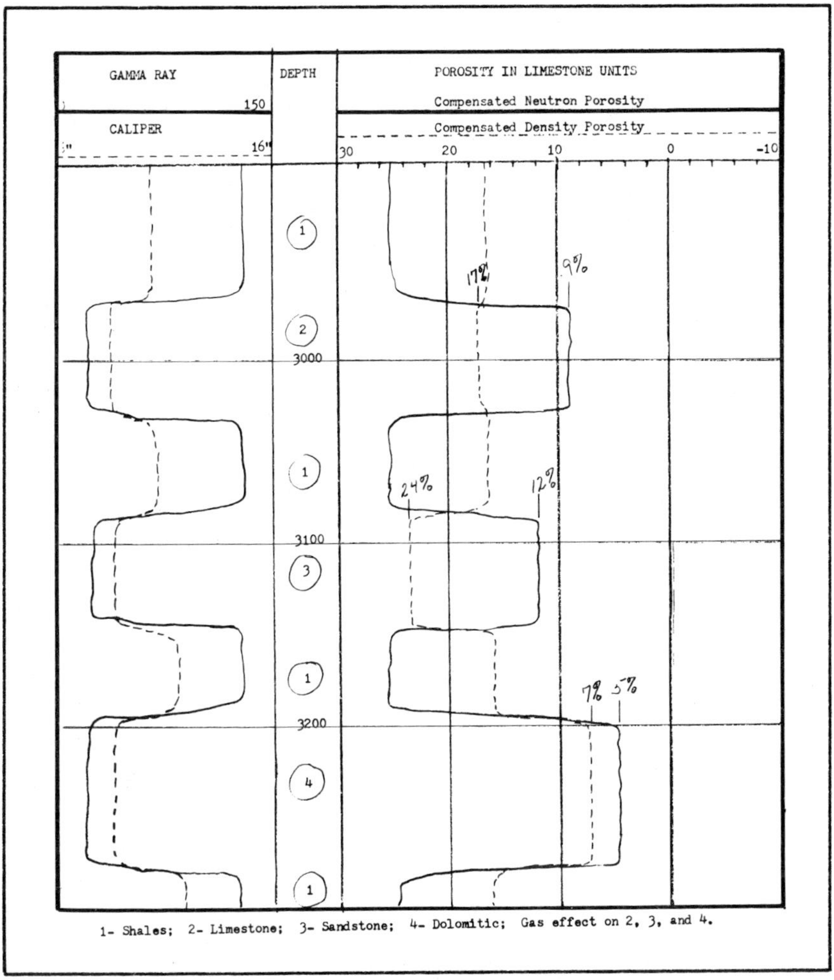

Schematic Illustration Of Lithologic And Gas Effects On Responses Of Neutron and Density Logs

Figure 74

log scaled in porosity for a particular rock type. Variations in the responses of the two devices identify changes in rock type and/or effects of gas. In Figure 73 we have an idealized illustration of this combination of logs scaled in limestone porosity units, with a Gamma Ray curve.

In Figure 73 Intervals number 1 are shales. The Gamma Ray level is high and the Neutron Log indicates porosity greater than the Density Log.

In Interval number 2 the Gamma Ray level is low with the Neutron Log and the Density Log indicating equal porosity values. This is the response in limestone.

In Interval number 3 the Gamma Ray level is low with the Density Log indicating porosity greater than the Neutron Log. This means the weight of the rock being measured is lighter than limestone, and this response is usual in sandstone or chert. The difference between the curves will be in the order of 6 - 8 porosity units.

In Interval number 4 the Gamma Ray level is low with the Density Log indicating porosity less than the Neutron Log. This means the weight of the rock being measured is heavier than limestone. Dolomite causes this response and the difference will vary depending on the extent of the chemical change from limestone to dolomite.

The responses indicated in Figure 73 are those expected when only liquid is involved. If there is gas in the porous spaces the logs will respond to the presence of that gas. The effect is to cause the Density Log to indicate greater porosity and the Neutron Log to indicate lesser porosity. The amount of this effect can be rather great depending on the gas saturation. Figure 74 is an example of gas effect; this is the same as Figure 73 except now the zones have gas in the porous spaces. The logs respond to that gas as shown in the figure. The shales (Interval number 1) are unaffected. Interval number 2 (Limestone) now appears to have sandstone characteristics. Interval number 3 (Sandstone) still has sandstone characteristics only more separation. Interval number 4 (Dolomitic) has lost its typical dolomitic response and can appear as a limestone or even sandstone, depending on the extent of dolomitization and amount of gas involved.

If we had either of these logs alone, there would be no way to know whether or not we had only lithology (rock type) effect, or both lithology and gas effects. Proper analysis in this case would be most difficult. This is the reason that the use of these logs in combination has become so widely used.

The analysis of these logs involves the use of a technique called crossplotting. A chart was developed to use in this procedure. Figures 73 - A and 74 - A illustrate the crossplotting of Figures 73 and 74. The porosity values indicated in Figure 73 are entered on their appropriate scales in Figure 73 - A, and lines are extended into the chart until they intercept. For example, both curves indicate 14% porosity in Interval number 2. Find 14% on the Neutron scale (bottom of chart) and strike a line upward. Find 14% on the Density scale (right side of chart) and strike a line left until it intercepts the line upward. Notice they intercept on the Limestone diagonal line at 14% porosity. The two porosity values 15 and 22% in Interval number 3 crossplot to the Sandstone diagonal line at 19.5% porosity. The Interval number 4 values of 3 and 11% crossplot about halfway between the Limestone and Dolomite diagonals. This means the limestone is approximately 50% dolomitized. Connecting points of equal porosity on each diagonal indicates porosity is 7%.

Using the same procedure on Figure 74 - A for the porosity values in Figure 74 we find the intercept points fall differently. Interval number 4 values of 9 and 17% intercept just above the sandstone diagonal line. It is important to discuss this point before we make the next crossplot. At this moment we may not know what type of rock is involved and whether or not there is a gas effect, but notice that the intercept falls at about 13.5% porosity. So regardless of lithology or gas, we have arrived at a porosity value very close to the proper value for this reservoir. However, if we know the geological column, and know what type of formation we are dealing with, we can refine our crossplot with that knowledge. Suppose this is a known sandstone, our intercept point is very close to the sandstone diagonal, so we have arrived at the proper analysis. If we should know that

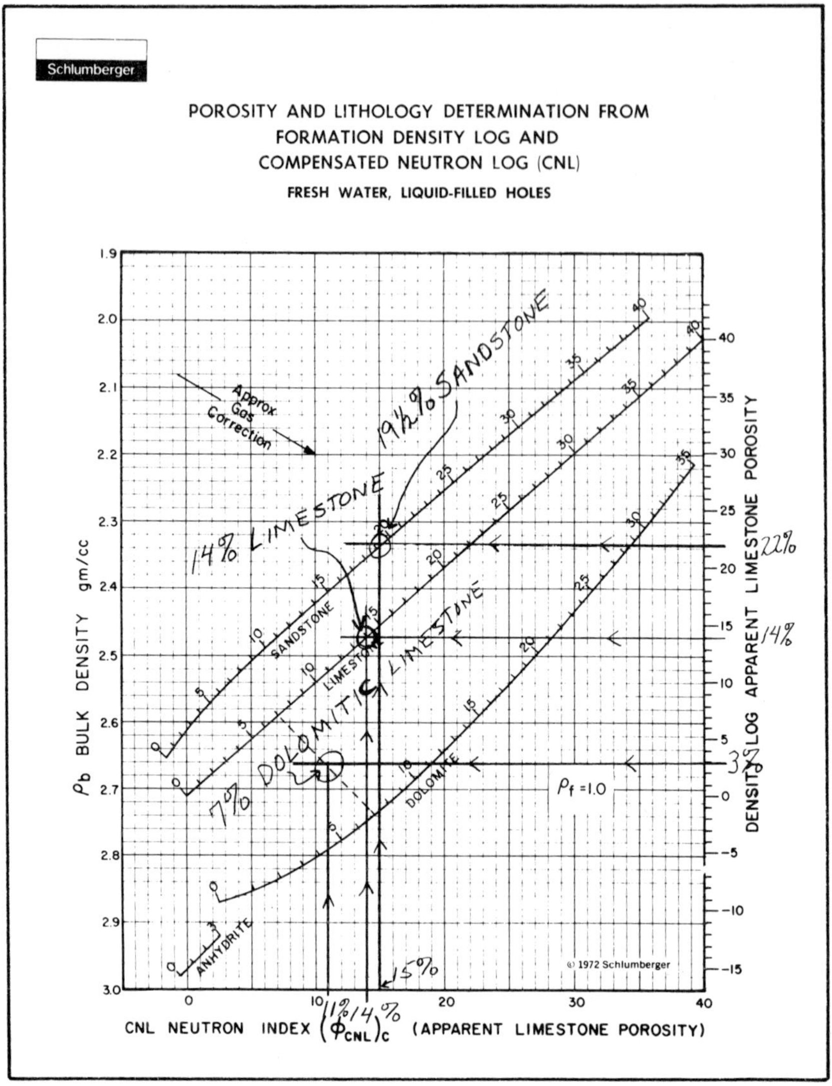

Crossplot Chart for Figure 73
Figure 73A

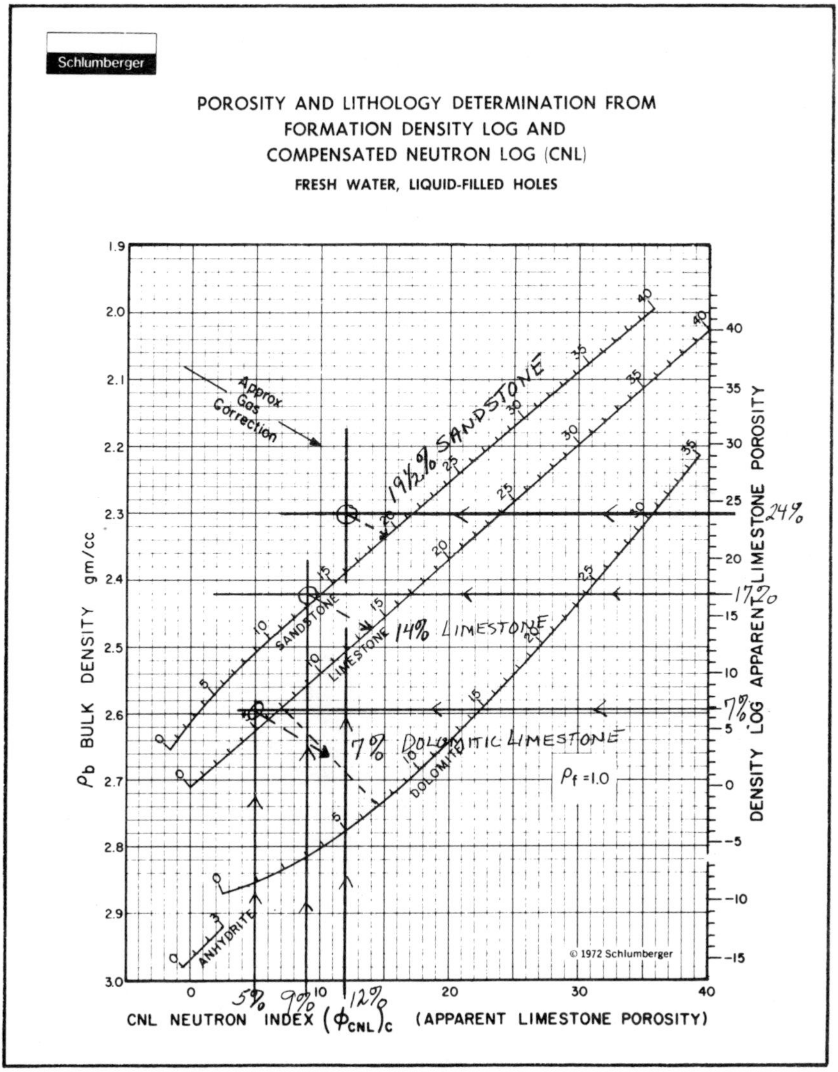

Crossplot Chart for Figure 74
Figure 74A

this is a limestone, we must get our point of intercept back to the limestone diagonal. The effect of gas on the crossplot is to cause the intercept to fall northwest (upwards and left) of where it should. The gas correction direction is shown by the arrow in the upper left corner of the chart. By moving our intercept point in this direction to the limestone diagonal we find 14% porosity.

Interval number 3 porosity values crossplot northwest of the sandstone diagonal. Here is obviously a gas effect, and if we do not know lithology we won't know whether to correct to limestone or sandstone (it is very unlikely that we would have that great an effect to have to go to the dolomite diagonal). But in either event, we would get 19 to 20% porosity. If we know the geological column and that this interval is a sandstone, then move the intercept in the arrow direction to the sandstone diagonal and find 19.5% porosity.

Interval number 4 crossplots just above the limestone diagonal at 6% porosity. Again, if we know nothing about lithology or gas we might consider this as a slightly gas effected limestone point. If we correct all the way to the dolomite diagonal we will find 8% porosity. So, with no knowledge, we have found porosity is in the order of 6 - 8%. Knowledge that this is a dolomitic limestone lets us refine the porosity to 7%.

So you can see that the crossplotting technique is a tremendous aid in analyzing the logs. Fortunately, we very often have some help in that the intervals we are working on are not as uniform as the idealized ones in Figures 73 and 74.

Figure 75 is an actual log with the Neutron-Density combination. The crossplot is shown in Figure 75 - A. Just looking at the log indicates something happens in Interval number 3 as compared to Intervals number 1 and 2. If we crossplot Interval number 1 and enter porosity values of 14 and 19% into the chart, we find the intercept just below the sandstone diagonal at 17% porosity. Interval number 2 porosity values of 13.5 and 18.5% intercept just below the sandstone diagonal at 16.5% porosity. Now Interval number 3 values of 7 and 25% intercept northwest of the sandstone

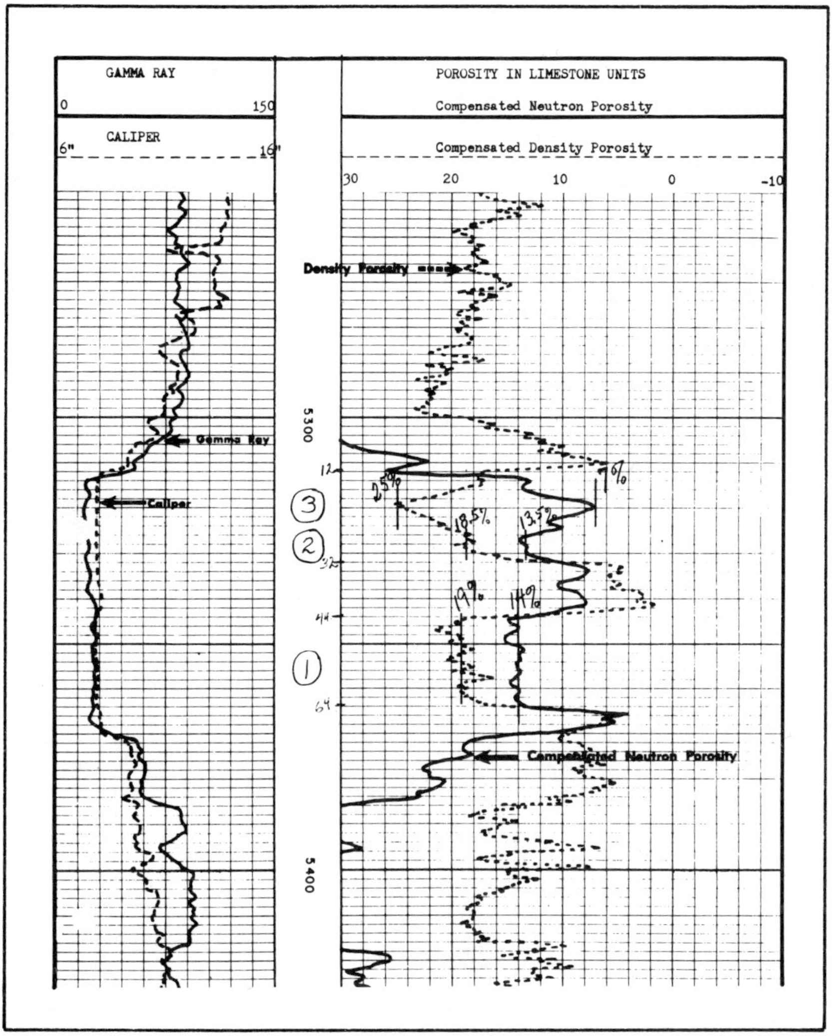

Compensated Neutron Log And Compensated Density Log
Through A Sandstone Section
Figure 75

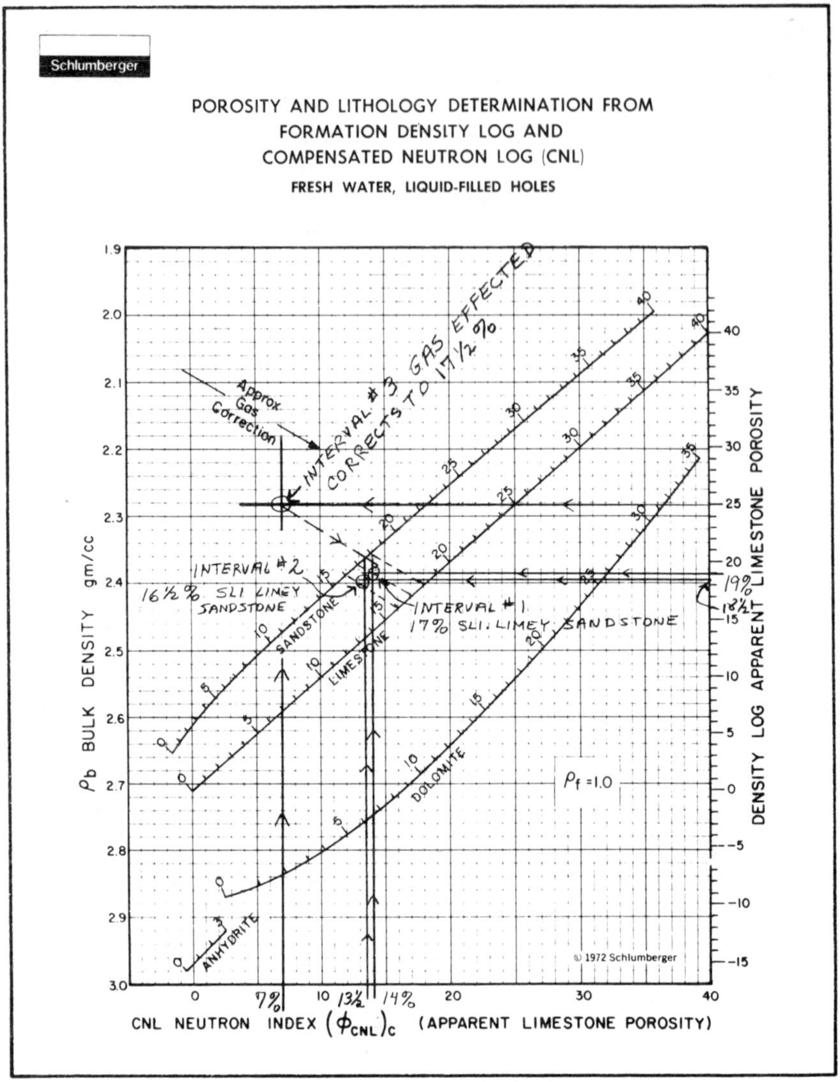

Crossplot Chart for Figure 75

Figure 75A

diagonal. This obviously is a gas effect. At this time we could conclude that maybe all three points are gas effected and we might correct to the limestone diagonal. If we do we will find porosity still falls between 16 and 18%. However, by knowing the geological column we know this is a sandstone section. With this in mind we can now conclude that there is a gas effect only in the Interval number 3, and since Intervals number 2 and 3 are adjoining there is a gas cap in this zone. We might also conclude that this sandstone is slightly limey or has a calcareous cementing material in it.

From what we have done so far, one can see that modern logging techniques can help provide answers not available before. Comparison of analyses of these logs and core data have shown a remarkable degree of accuracy and dependability in determining rock type and porosity values. Whether or not data derived from work with cores is always accurate is debatable. However, it is an accepted criteria and comparison of the two methods is routinely made.

Now that we have found that the porosity of the reservoir can be determined we will proceed to the next step, which is to determine the water saturation of that porosity. When we have a proper value for water saturation we can determine hydrocarbon saturation. The accepted procedure is to consider that the porosity is filled with something, and if part of it is filled with water the rest must be filled with hydrocarbons. Consequently, hydrocarbon saturation is equal to 1 minus water saturation.

The basic relationship for determination of water saturation is Mr. Archie's saturation formula, which is $\mathbf{Sw} = \sqrt{\frac{\mathbf{F \times Rw}}{\mathbf{Rt}}}$. F of course is formation resistivity factor, which we obtain from porosity. Rw is the resistivity of the formation water contained in the porous spaces. Rt is the true resistivity of the formation in its uncontaminated state, which is beyond the invaded zone. Rt is determined from the deep investigating curve of the resistivity logs. As we have discussed previously, there were many complications in determining Rt from logs prior to the advent of focused logs.

The first of the focused logs was the Laterolog (Guard Log) which was, and is still, used primarily in geological provinces where salt contamination of the mud system is a problem. If in the process of drilling a well salt beds are encountered, the salt is dissolved and the mud becomes very salty. This causes the mud column to become very conductive. Electricity, being lazy, follows the path of least resistance so the logging currents would flow up and down the bore hole rather than out into the formations as they should in order to make proper resistivity measurements. This would make the non-focused logs practically useless.

The Laterolog is a method in which logging current is forced to flow into the formations by using a focusing arrangement of the electrode system. Since the mud is conductive the filtrate that invades the porous permeable formation creates an invaded zone that is less resistive than Rt in an oil or gas bearing formation. This low resistivity invaded zone does not present much of a barrier to the measurement of true resistivity (Rt). Consequently, quantitative analysis of logs in salt provinces became a reality.

§ 4.04 Induction Logs

In areas where salt was no problem and wells were drilled with fresh muds, the problems, as discussed previously, were due to bed thickness, bed resistivity, and invasion. The Induction Logs were developed to overcome these problems.

The first Induction Log presentation was a combination of the old style Electrical Log and the Induction Log. The shallow investigating curve was the Electrical Log short normal, generally 16″ spacing, and the deep investigating curve was the Induction Log. The Spontaneous Potential was kept from the Electrical Log also. Since there are only two depths of investigation it is difficult to determine whether or not invasion is deep enough to cause problems.

Further development brought forth the Dual Induction Log, which added a medium investigating Induction curve and replaced the shallow curve with a specially designed Laterolog. This Laterolog has an electrode arrangement that restricts its measurement to very close to the bore hole so it will measure predominately the resistivity of the invaded zone. This log also has retained the Spontaneous Potential curve. The three focused curves permit very accurate analyses to be made.

Let us now look at some logs and determine some water saturations. In Figure 76 we have an Induction Electrical Log through the limestone section shown in Figures 69 and 72. The porous interval (from the Sonic Log and the Density Log) is from 6260′ to 6273′, so this is where we want to determine Rt. The deep curve (dashed) reads 32 ohm-meters through this interval. With no other information we will accept this value as Rt.

The next thing we need to know is the Rw. The best source for water resistivities will be from catalogs of measured samples, if possible. There are many of these available from every geological province. Other sources for Rw would be from the S. P. curve and from nearby water sands. However, the catalogs are the best.

Rw for our example is .04 ohm-meter at formation temperature of 140°.

F is determined from the porosity using the relationship most suitable for carbonate formations. This is $\mathbf{F} = \frac{1}{\emptyset^2}$. From the Sonic Log we found porosity = 13.5 %. $\mathbf{F} = \frac{1}{.135^2} = \mathbf{55}$. From the Density Log we found porosity = 20 to 23 %. $\mathbf{F} = \frac{1}{.20^2} = \mathbf{25}$ and $\frac{1}{.23^2} = \mathbf{19}$. This is quite a variation and could present a problem. Before we draw any conclusions let's determine the water saturation for each value. Rather than go through the mathematics for each set of numbers we can use a chart available to do that for us. The chart is shown in Figure 77. Enter this chart at the left on the stem labeled Rw at .04. Find F = 55, 25, and 19 on the second stem labeled $\emptyset$% F_r. Draw a line from .04 through each point for F to the stem labeled Ro. This gives us the resistivity expected if the formation

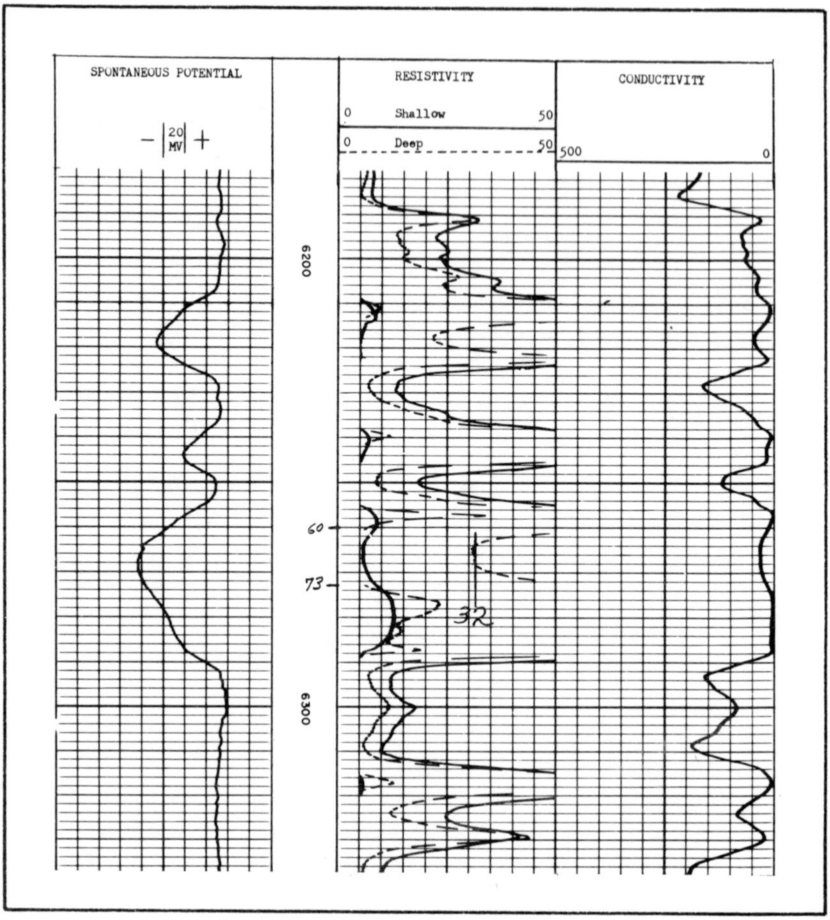

Induction Electrical Log Through A Limestone Section
Figure 76

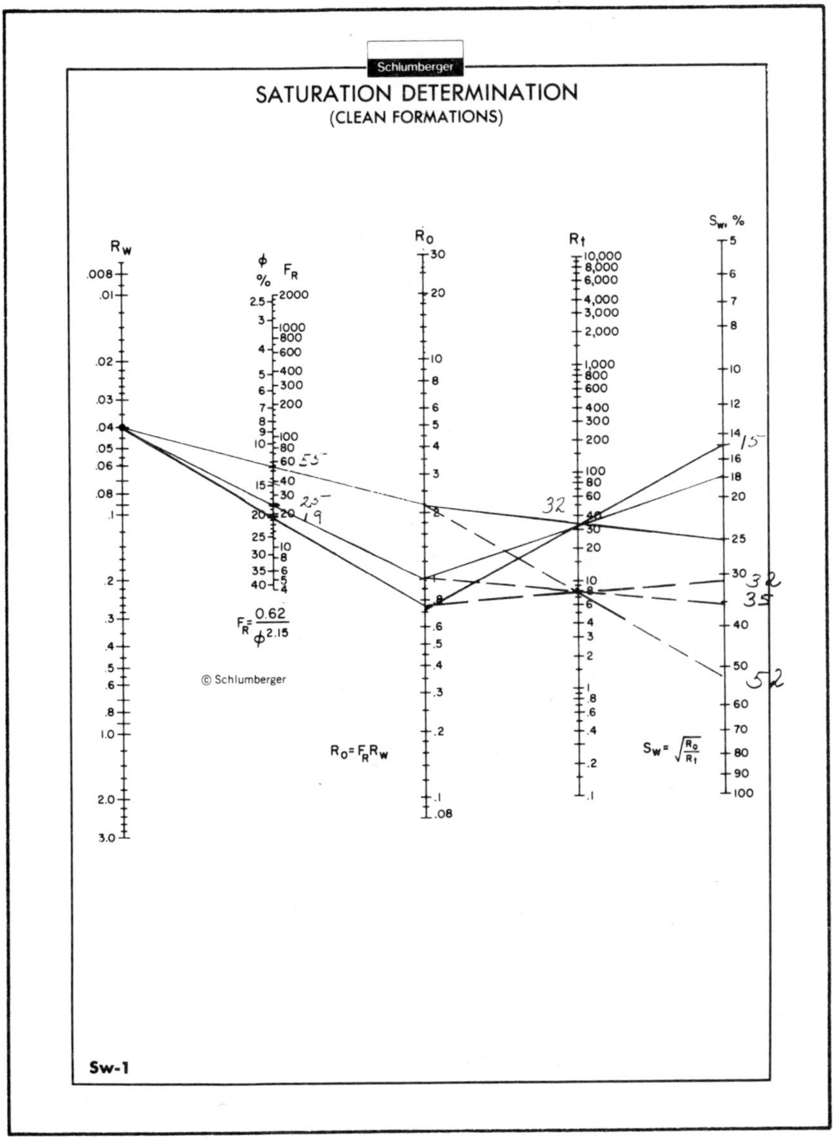

Schlumberger
SATURATION DETERMINATION
(CLEAN FORMATIONS)
R_w
ϕ %
F_R
R_o
R_t
S_w, %
$F_R = \frac{0.62}{\phi^{2.15}}$
© Schlumberger
$R_o = F_R R_w$
$S_w = \sqrt{\frac{R_o}{R_t}}$
Sw-1

Figure 77

were 100 % water filled. Find 32 on the stem labeled Rt and draw a line from each Ro through 32 to the stem labeled Sw. We have solved the formula $\mathbf{Sw} = \sqrt{\frac{\mathbf{F \times Rw}}{\mathbf{Rt}}}$ for each value of F. Sw ranges from 15 to 25 %. At these low values one would anticipate hydrocarbon production from this interval. It was completed and made a good well.

In this instance the broad range of Sw did not cause the interval to be condemned. The saturation was so low that misinterpretation was unlikely. We were fortunate enough to have logs that permitted a proper solution to be made. We could determine that there was secondary porosity involved and that the Sonic Log porosity was too low; consequently the Sw derived using that value was too high.

By way of explanation, let's digress for a moment and discuss what is commonly called "Critical Water Saturation". As was stated previously, when hydrocarbons accumulate in a reservoir they displace some of the original connate water. If enough hydrocarbons accumulate so that all the movable water has been displaced, we have water saturation (Sw) at the irreducible value. No more can move. A reservoir at this saturation would produce nothing but hydrocarbons.

If the reservoir is at Sw greater than irreducible, then there is some water that is free to move. How much will move depends on a rock characteristic called relative permeability. This is a measurement of the ability of each fluid (oil or water) to flow in the reservoir. At irreducible Sw only oil will flow, but as Sw increases the relative permeability to water (Kw) increases and the relative permeability to oil (Ko) decreases. This means that at Sw greater than irreducible we can expect some water cut with the production. At some Sw value more water than oil will flow.

Figure 78 is a set of typical relative permeability curves. Notice that at about 15 % Sw permeability to water (Kw) is 0 and permeability to oil (Ko) is .9, which means only oil will flow from this reservoir at this saturation. As Sw increases so does the permeability to water; permeability to oil decreases so water production increases. Somewhere around 60% Sw Kw starts

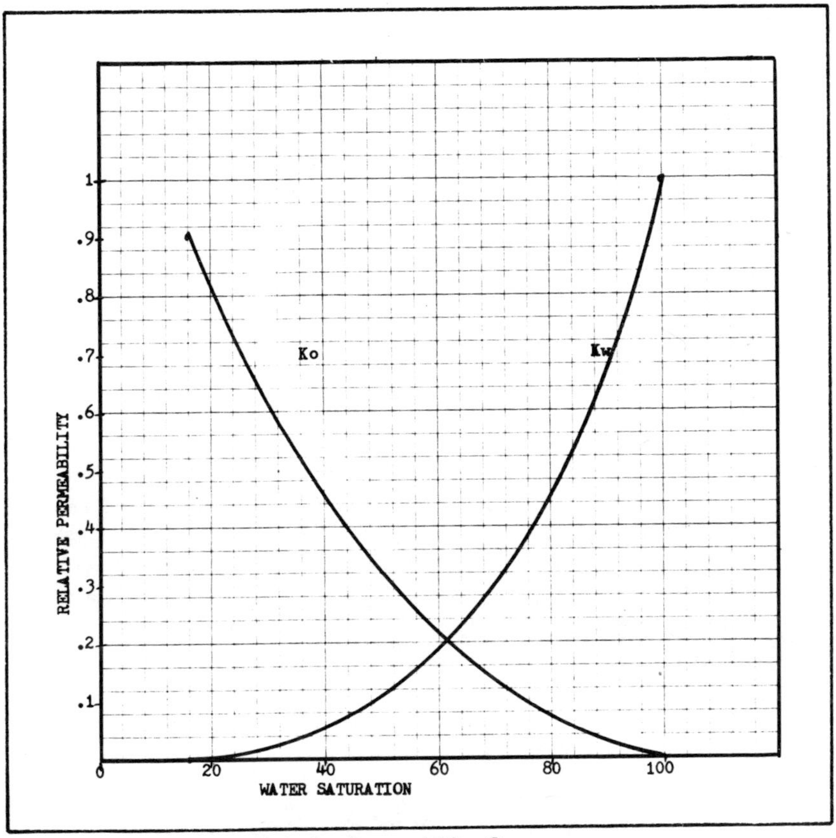

Relative Permeability Curves
Figure 78

exceeding Ko so predominately water will be produced.

Generally, irreducible Sw will be at the upper structural limit of the reservoir, and as we go down dip there will be more water and less oil until we reach a point where there is only water. The area between the upper limit and the water table is called the transition zone. This is caused by capillary pressures in the reservoir, and each rock type will have its own characteristics as to how long the transition zone is. Low porosity and permeability rocks may have transition zones up to many hundreds of feet long, while high porosity and permeability rocks may have short transition zones of only a few feet.

Now back to our limestone section in Figure 76. The Ko - Kw curves for this reservoir could be similar to those in Figure 78 with irreducible Sw between 15 and 20 %. So even though we found Sw = 25 % using the porosity and F from the Sonic Log, the very worst we should expect would be a slight water cut with the production. Now suppose this reservoir were in the transition zone (Sw greater than irreducible) and true resistivity from the deep curve was around 8 ohm-meters. Using the same Rw and F's as before we would find Sw's between 30 and 35 % and 52 % as shown by the dashed lines on Figure 77. The correct porosity from the Compensated Density Log would give Sw that would indicate production of oil with possibly some water cut. The too low porosity from the Sonic Log would give Sw that would indicate the zone is near to where water with some oil would be produced and could cause the zone to be condemned. Correct porosity is very important to properly evaluate a formation.

Figure 79 is the Dual Induction - Laterolog through the sandstone illustrated by the Compensated Neutron - Compensated Density Log in Figure 75. We will determine the water saturation in this sandstone.

Two intervals of porosity are indicated: one from 5312′ to 32′ and the other from 5344′ to 64′. If we study the resistivity values in these intervals we will see that the two induction curves are reading the same and somewhat lower than the laterolog. This indicates that invasion is probably not very deep so that both induction curves are Rt. So if we pick values

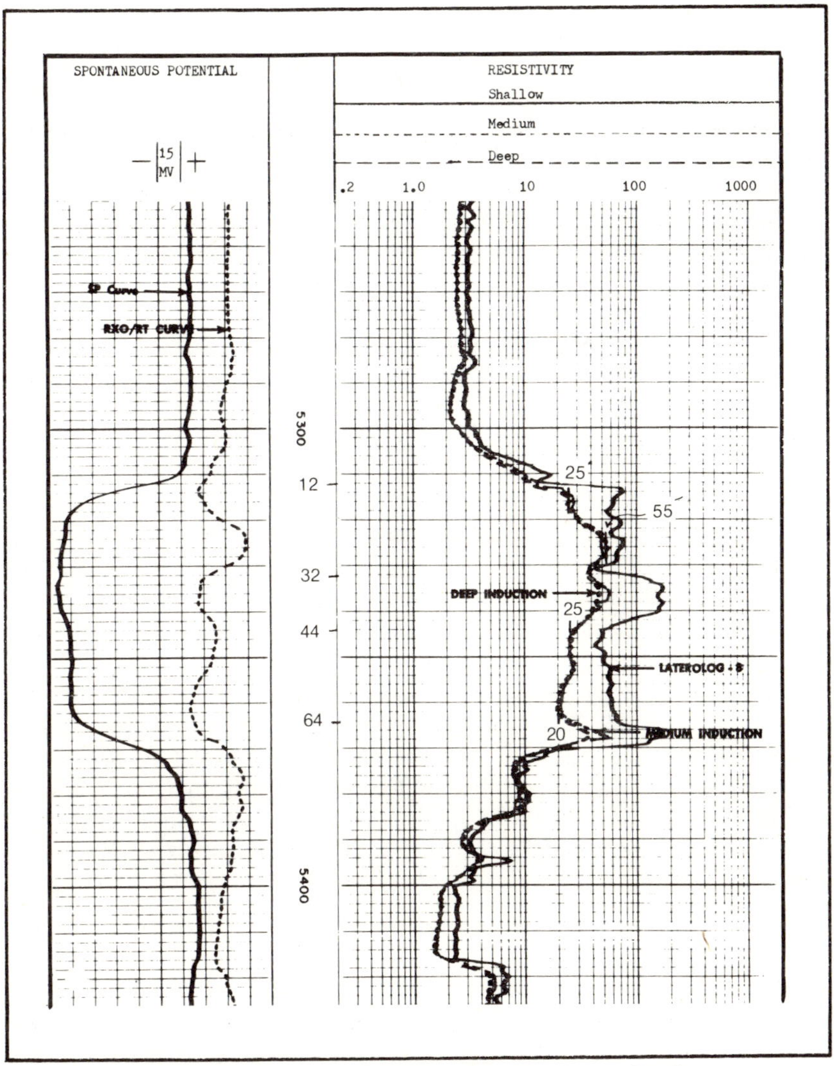

Dual Induction—Laterolog Through A Sandstone Section
Figure 79

corresponding to our porosity intervals we will be ready to determine water saturation except for an Rw value. From our catalog of Rw's we find an average value for this formation is .05 at formation temperature of 130°.

The upper porous zone has porosity = 17.5 % in interval 3 (Figure 75) and 16.5 % in interval 2. Where porosity = 17.5 % Rt = 25 and where porosity = 16.5 % Rt = 55. In the lower porosity interval (no. 1) porosity = 17 % and Rt varies from 20 to 25.

Figure 80 is another copy of the chart used to determine Sw. Find .05 on the Rw stem, and since this is a sandstone we will use the formula for F as shown under the ØFR stem and find the porosity values 16.5, 17, and 17.5 on the Ø stem. Draw lines from .05 through each porosity value to the Ro stem. We will find Ro = 1.5 for porosity = 16.5%, 1.4 for Ro in the 17% zone, and 1.3 for Ro in the 17.5 % zone. Find Rt = 20, 25, and 55 on the Rt stem. Draw lines from each Ro through the corresponding Rt to the Sw stem and find that Sw varies from 16.5 to 27%. These values are low enough to indicate the probability of hydrocarbon production from the two porous intervals. The well was completed for a gas well from perforations in the top part of the upper porous interval.

Figure 79 has an additional curve in the S.P. track. The dashed curve is labeled Rxo/Rt. This curve is made by a computer in the surface equipment and is useful as a "Quick Look" reference to identify formations of interest. These formations are then studied in detail by the log analyst.

The curve is the result of the following manipulation. If we consider only water bearing formations, then the resistivity of the flushed zone Rxo = F x Rmf and the true resistivity Rt = F x Rw. If we divide the first by the second

$$\frac{\mathbf{Rxo = F \times Rmf}}{\mathbf{Rt = F \times Rw}}$$

we find the F's cancel and Rxo/Rt = Rmf/Rw. Remember the S. P. is a result of the Rmf/Rw ratio and Rxo and Rt are measured by the shallow and deep resistivity curves, respectively. So in water bearing formations the two curves will tend to "track" each other as shown in Figure 81. If there are hydrocarbons involved the resistivities change considerably. Rt by the total oil or gas and Rxo by the residual oil or gas, and the ratio Rxo/Rt becomes much smaller. This causes the two curves to depart from

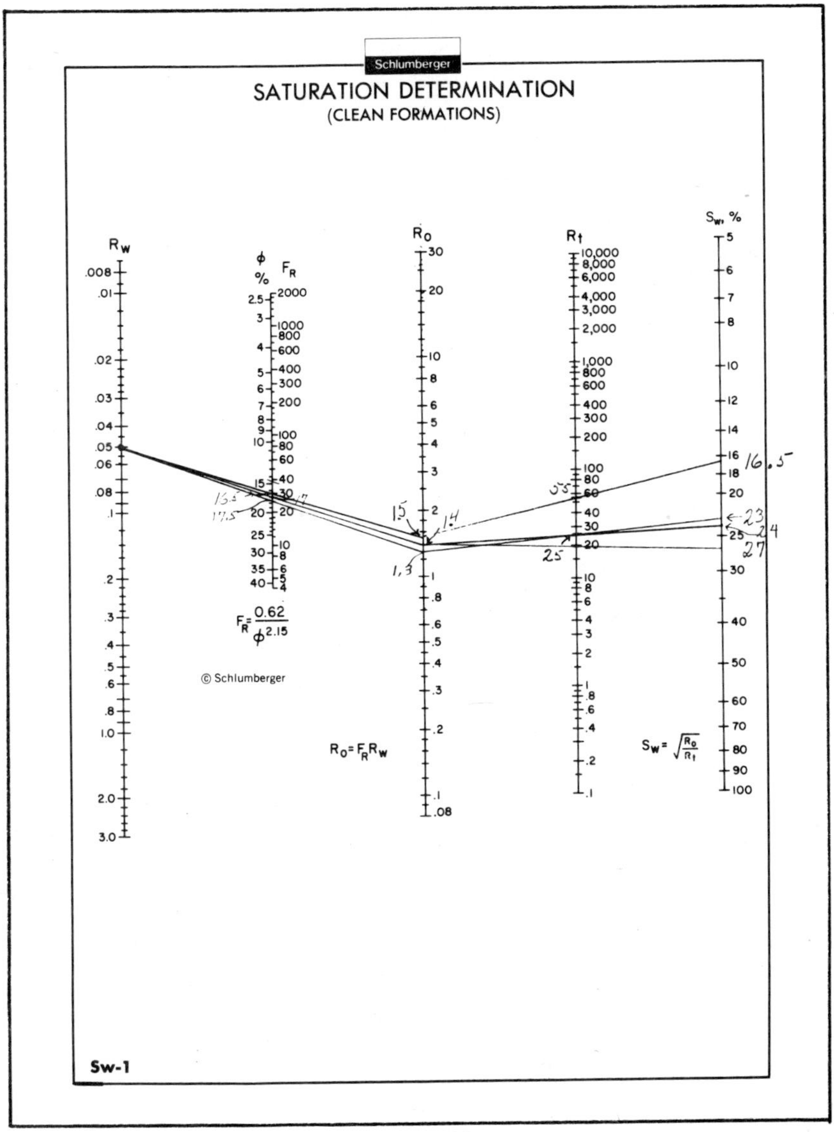
Schlumberger
SATURATION DETERMINATION
(CLEAN FORMATIONS)
Rw
φ %
FR
Ro
Rt
Sw, %
FR = 0.62 / φ^2.15
© Schlumberger
Ro = FR Rw
Sw = √(Ro/Rt)
Sw-1

Figure 80

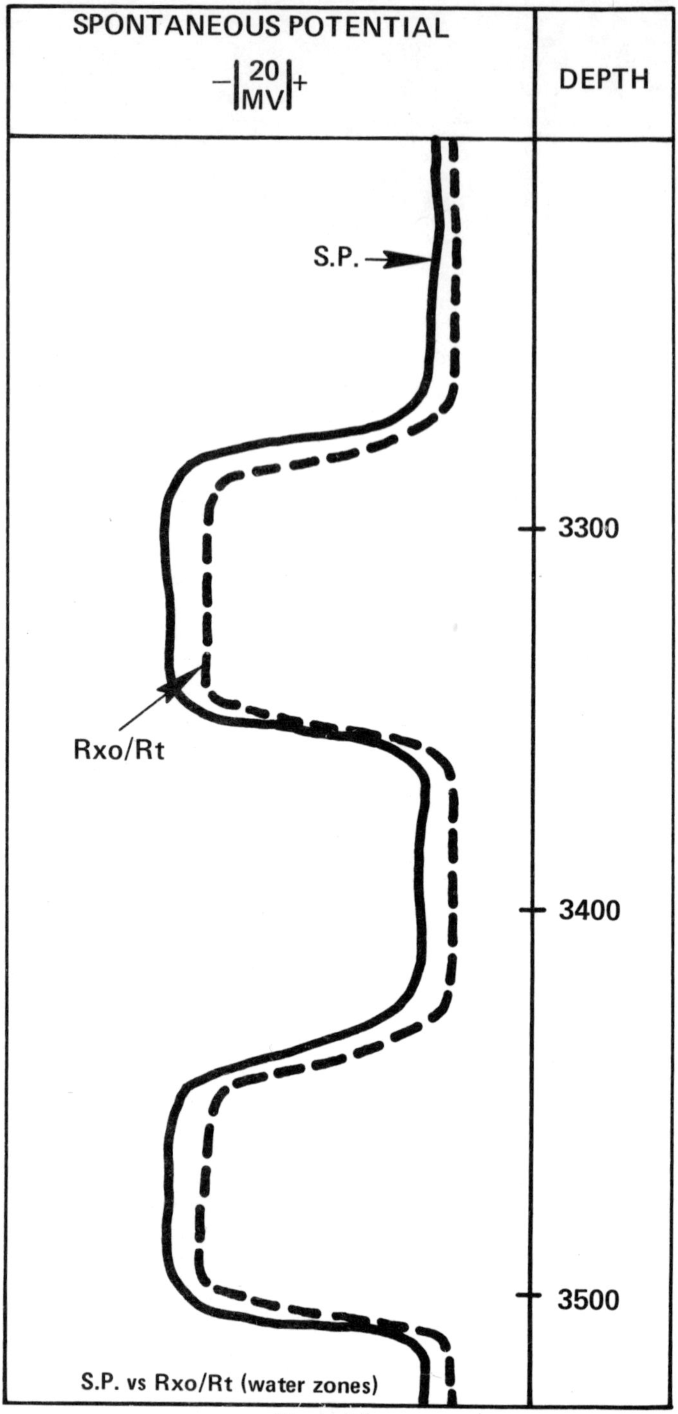
SPONTANEOUS POTENTIAL
20 MV
DEPTH
S.P.
3300
Rxo/Rt
3400
3500
S.P. vs Rxo/Rt (water zones)

Figure 81

each other as seen on our example in Figure 79. There is nothing quantitative about this relationship; it is merely to call your attention to formations of interest.

Most of the logging companies now have computers as part of the surface equipment in their field units. This permits several types of "Quick Look" techniques to be used. The quick look techniques are not intended to replace detailed analyses, but to attract your attention to zones that might have possibilities. There is a certain degree of optimism incorporated into the techniques. The general feeling is that it is better to cause some investigation or testing to be done rather than pass up a possible pay zone.

Some of these techniques require some drafting by the logging engineer while others may be recorded directly with the surface equipment. If, for example, a combination of resistivity curves and a porosity log are recorded simultaneously, the field computer can solve for apparent water resistivity. This is called the Rwa method of quick look. The reasoning here is that, since Ro = F x Rw and Rt = Ro in a water sand, if the deep resistivity Rt is continuously divided by F the solution is Rw if the zone is water filled. If there are hydrocarbons involved then Rt is greater than Ro and the resulting Rwa is greater than the true Rw for the zone. This is all possible because the computers can solve the porosity log for porosity and simultaneously derive F, which is divided into Rt continuously as the logs are run. The Rwa curve is recorded in the S. P. track on a sensitive scale. The scale is generally 0 to 1.0 for full track deflection with 0 at the left side of the log. Deflections to the right in porous intervals are indications of possible hydrocarbons. Figure 82 is an idealized illustration of this procedure. The S. P. is in the left track, the shallow investigation curve and the deep investigation curve are in the center track using the first two cycles of the logarithmic scale, and the Sonic Log in the right track. The Rwa curve is in the S. P. track as a dashed curve.

A method involving drafting by the logging engineer is one in which the logs are recorded on separate trips in the hole. When the porosity log is run the computer solves for porosity and simultaneously derives F. The F curve

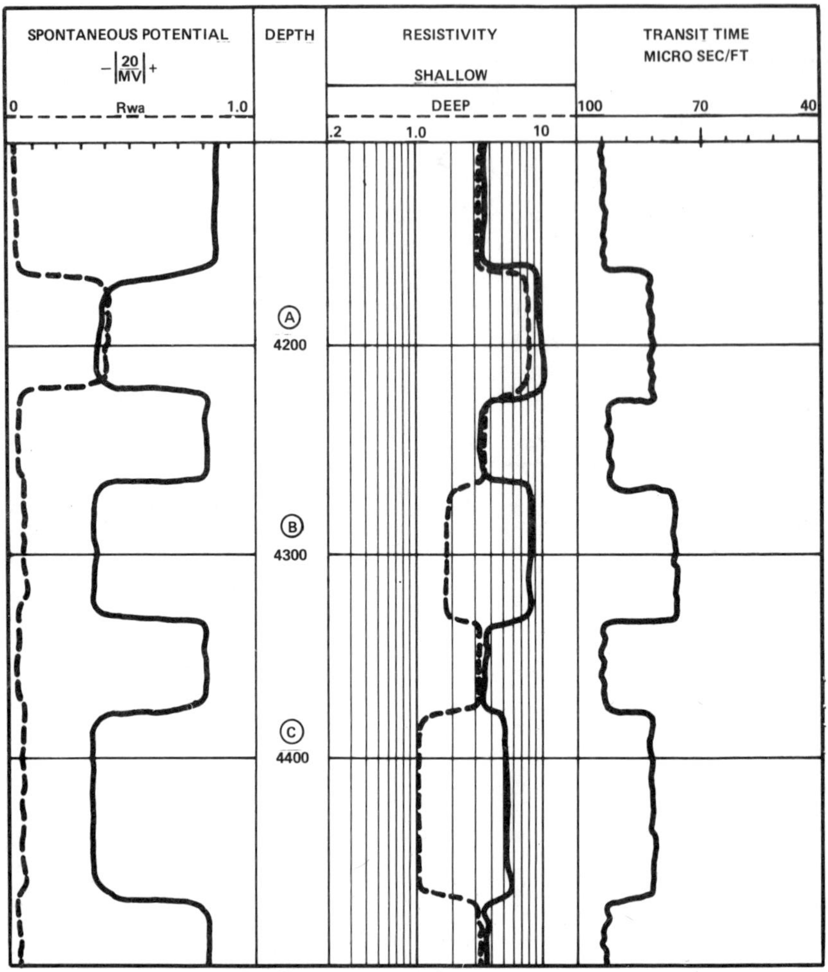

Ⓑ & Ⓒ are Water Filled   Ⓐ Possible Hydrocarbons

Figure 82

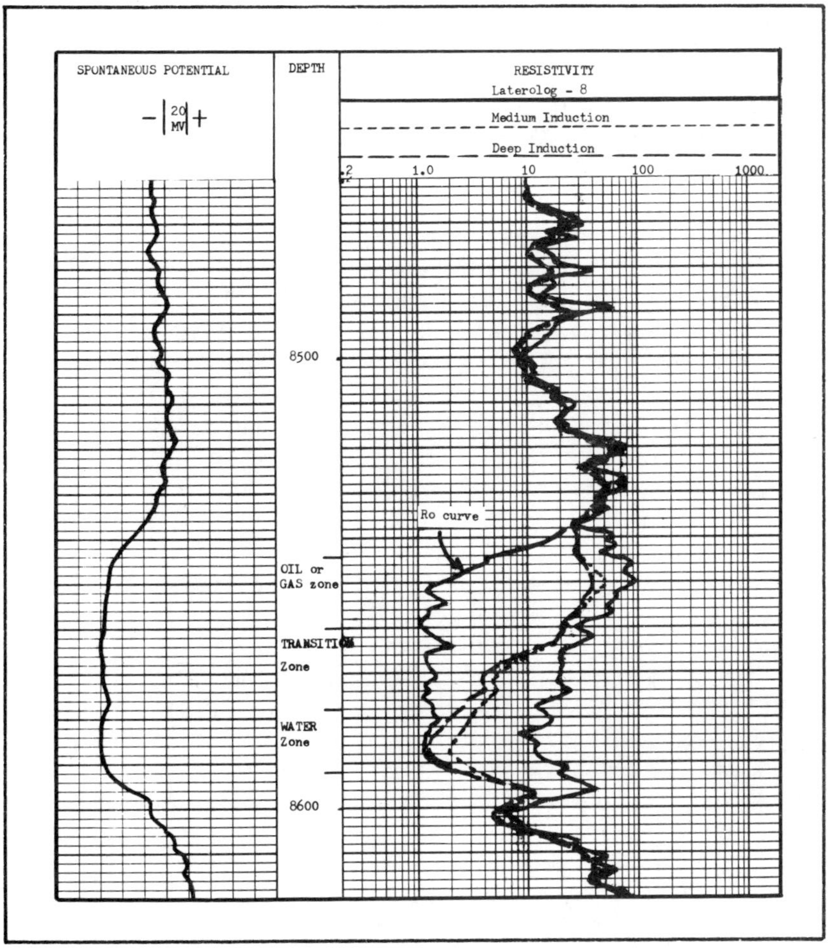

Ro Curve From F Curve Overlay
Figure 83

is recorded on a logarithmic scale so it is compatible with the resistivity curves also recorded on logarithmic scales. The logarithmic scales permit the F curve to be manipulated with the resistivity curves in a manner similar to a slide rule. The engineer places the film for the F curve under the film for the resistivity curves, and slides it until he makes the multiplication F x Rw. This, of course, equals Ro for the intervals of interest. He then drafts the F curve onto the resistivity film with India ink. If the F curve and the deep resistivity curve are coincidental the zone is water filled, but if the resistivity curve departs to the right from the F curve then hydrocarbons are involved. An illustration of this procedure is shown in Figure 83.

There are other quick look techniques which utilize similar procedures, but the Rxo/Rt, Rwa, and Ro presentations are the ones most often seen.

When the intervals of interest have been isolated they are then analyzed in detail. This analysis usually results in the tabulation of two numbers, porosity and water saturation. The important question then is: Will this zone produce hydrocarbons at the numbers determined? The answer often lies in the knowledge gained from experience in the area. One might know that a particular sandstone is generally a fine grained silty sand and will produce relatively water free hydrocarbons at Sw as high as 60 %, so long as porosity is in the average range of 15 %. Yet another sandstone that is course grained with high permeability would produce nothing but water with that high saturation. Such a sandstone might need Sw in the order of 35 to 40 % to produce water free. Sometimes it takes a long time to gain this necessary experience.

An analysis technique that helps overcome this difficulty involves plotting values of porosity and water saturation on linear graph paper. If a reservoir has Sw at the irreducible value the plotted points will form a coherent curve at a constant value. When Sw exceeds the irreducible saturation the points will no longer fall on the constant value curve. This procedure is described in detail in a paper available from Schlumberger Well Services. Figure 84 shows an example of this technique. Points 1 through 8 form the coherent curve (Ø x Sw always give the same product);

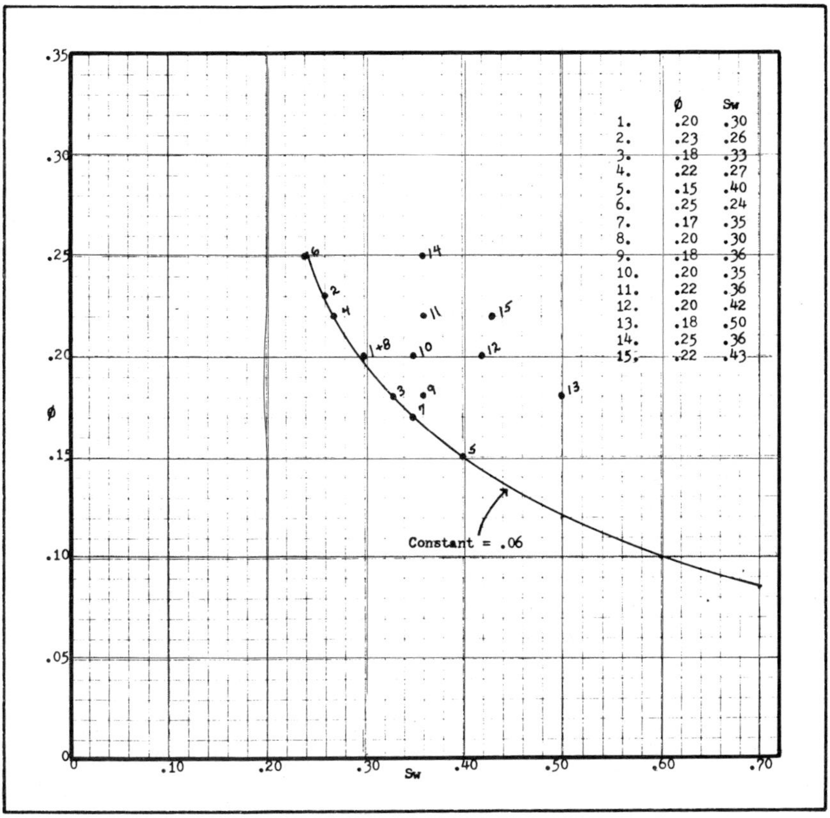

Plot Of Sw vs Ø
Figure 84

points 9 and below fall to the right of the curve and indicate that Sw has become higher than irreducible saturation. Water free hydrocarbons should be produced above point 9; below point 9 is transition zone and the amount of water produced will depend on the relative permeabilities.

The analysis techniques that have been presented in this chapter are just some of the ones in use by log analysts. There are always exceptions to the norm in nature and unusual conditions are always being found. Generally these conditions involve formations that exhibit lower than expected resistivities yet produce hydrocarbons. Shaley sands are notorious for this, and there are many specialized techniques for shaley sand analysis. Some carbonate reservoirs also fall in this category. They, too, require special techniques in order to make a proper analysis. There are formations that contain conductive minerals that affect the measured resistivities and cause serious problems in arriving at the proper conclusions.

Many coordinated efforts have been involved in developing procedures that help solve these problem conditions. Data from cores, logs, and production tests, along with a tremendous amount of laboratory work, has gone into deriving these solutions. The effort is justified because much production has been found as a result of this work.

There will be no attempt to discuss these specialized techniques here, but it is necessary to know that they exist. They have been reported in the literature and papers describing them in detail are available from many sources, such as logging companies and professional societies, should one wish to pursue the matter.

Before we leave the chapter on analysis and interpretation of the logs, it must be reported that the logging companies have advanced computer programs available that go beyond the field procedures. When the logs are being run they may be recorded in digital form on tape at the same time they are recorded in analog form on film. The tapes are sent to computer centers where they are processed through very sophisticated programs, and the results are presented on film compatible with the analog film from the field. The two are then readily correlatable for further study.

Chapter Five

SELECTION OF LOGGING PROGRAMS

The selection of a logging program is a very important part of setting up the prognosis of a well, especially when one considers the great expense that can be caused by improper or insufficient information, and the loss of revenue by plugging a potentially productive well.

There are logs and combinations of logs which, when properly applied, will provide answers in even the most complicated situations.

Of course, every well will not need a complex program, because the answers required are relatively simple, and the information available prior to drilling the well may be great. One example is in-fill drilling in a field where there is much geologic and production data and only one or two possible pay zones. Such a well will not demand the logging program that one in a wildcat area with much less geologic or production data and many possible pay zones ranging from sandstones to complex lithologies would require. In either case, consideration must be given to the answers required for successful completion of the well, and logs selected that will provide those answers.

§ 5.01 Logging Program Considerations

Some of the basic considerations necessary to select a logging program are as follows:

1. Bore hole fluid - Fresh mud, Salty mud, oil or oil base mud, empty hole (air or gas drilled hole).

2. Formations encountered - Sand shale sequence - high porosity sands, sand shale sequence - low to medium porosity sands, shaley sands, tight sands, carbonates, complex lithology problems.
3. Indications of gas.
4. That the logs selected will provide the information required from the drilling of the hole; that is, a complete permanent record of the hole, accurate determination of the storage space (porosity), accurate determination of hydrocarbon saturation (Sw), accurate determination of pay thickness, indications of permeability, accurate depth control, and correlation.
5. Additional geologic data - Dipmeter.

There are conditions that might be classified as "mechanical" that should be taken into account. A shallow well may be planned differently than a deep well, especially if the deep well is to be drilled to its total depth and logged only once.

A deep well prognosis may include one or more strings of intermediate casing. This could necessitate several logging runs, one for each time casing is to be set. There could even be different suites of logs in each case, depending on bore hole conditions and answers desired.

Sometimes pressure and/or temperature can become important in the selection of logs.

Selection of a suite of logs "because that is what was used in the offset well" can very often lead to difficulty. This is especially true if one is drilling in an area where offset logs may be of the older type. The reason given for this selection is quite often that comparison of the wells is more easily accomplished. Comparison may be accomplished, but evaluation may suffer.

Because there are so many different types of logs available it is virtually impossible for everyone to be familiar with all of them. If you do not have access to a log analyst then it is good to have a logging company representative, in whom you have confidence, to consult with in the selection of logs. Proper logging programs can vary in different geological provinces, yet fulfill the requirements.

Chapter Six

OTHER TYPES OF WELL LOGS

There are several other types of well logs that are widely used. The most common of these are: Drilling Time Logs, Sample Logs, and Mud Logs.

§ 6.01 Drilling Time Logs

Drillers were probably the first to learn that a change in drilling rate meant a change in the type of rock penetrated by the drill bit. Different lithologies would be recognized before the examinations of the cuttings confirmed them. Even though penetration rate may not be a scientific property of rock, and may not be catalogued quantitatively, it is certainly a qualitative characteristic.

The rate of penetration is the time required to drill a linear unit of depth of a geologic formation. It is expressed as a number of units of depth in a unit of time, for example "FT/HR". Drilling time is the amount of time required to drill a unit of depth, such as "MIN/FT".

Drilling time in Min/Ft is the most popular presentation because it is more exact and is useful where detailed lithologic information is desired. Also, it helps solve difficult correlation problems.

Drilling time logs very often correlate quite well with the Spontaneous Potential and Resistivity curves of electrical well logs.

§ 6.01(a) Applications. Some applications of drilling time logs are:

1- An aid to evaluating bit performance in various rocks.

2- A help in preventing drilling past possible pay zones. The driller can stop at a "drilling break" and circulate samples for examination.

3- A help in picking coring points.
4- Identification of thin shales or tight streaks in sandstones.
5- Correlation to electrical logs in other wells to help orient the operator as to the stratigraphic position of zones in the drilling well.
6- An aid to interpretation of other well logs.
7- Possibly one of the best reasons for the use of drilling time logs is their value to protect against the loss of geologic data in case the hole is lost or if hole conditions prevent running other types of logs.

§ 6.02 SAMPLE LOGS

Much subsurface geologic work depends on the collection of representative samples of the formations penetrated, accurate descriptions of these samples, and description plotting that allows the sections in different wells to be correlated.

The collection and examination of samples or well cuttings are highly organized and important techniques. In fact, in many areas well cuttings are the only reliable source of data concerning the formations drilled.

These samples are taken from the returning fluid stream at regular intervals. The sample interval is usually 5 or 10 feet, but may be as small as 1 foot or as large as 25 to 30 feet. The chief problems in collecting samples of rotary cuttings are prevention of contamination from upper beds, prevention of powdering of the sample, prevention of loss of the sample, prevention of separation of the sample, obtaining of correct depth measurement, and washing and drying the samples properly.

Contamination from upper beds is the chief problem of collecting rotary samples, especially where the hole is drilled with poor mud. The walls of the hole are poorly plastered and unstable, thus the samples contain large amounts of extra material. With properly treated mud the hole is well plastered and caves very little, so the percentage content of the sample may be taken as representative of the types of rock in the interval covered. This enables the geologist to follow, very closely, the lateral gradations in the section.

The powdering of the sample to a size too small to be examined effectively is largely due to poor mud, which allows regrinding of the sample by the bit and powdering by whipping of the drill pipe against the wall of the hole. Reverse circulation will overcome this difficulty because the mud velocity is greater and carries the cuttings to the surface. Reverse circulation is especially useful when drilling into low pressure formations by using oil as drilling fluid, because it prevents any of the fine oil-borne cuttings from plugging the pores of the formation. In regular circulation with oil, the low viscosity of the oil will not carry out the cuttings, and they are commonly reground to a fine putty-like mass that is useless for examination and has a plugging effect on low pressure pays. Of course, reverse circulation is not necessary in high pressure oil or gas pays, since the high pressures tend to increase the velocity of the circulating fluids and carry the cuttings out of the hole.

Loss of samples is due to two chief causes; lost circulation and blowouts. Both of these are primarily mud problems. Lost circulation is caused either by too high water loss in the mud or by excessively heavy mud which overcomes the formation pressure of a porous or fractured formation so that the mud enters the pores or fractures. The cuttings are then carried into the beds and may never be recovered. Blowouts are caused by unexpectedly high formation pressures or by carelessness in handling the mud. The cuttings are blown out of the hole and not recovered.

Separation of the course from the finer part of the sample by the upward movement of the circulating fluid is due to the use of mud with low viscosity and gel strength. In good mud the cuttings are held in suspension and there is little change in the relative position.

Obtaining proper depth measurement for rotary cuttings is another major problem. The depth of the well must be checked often, either by steel line or measuring the drill pipe under tension. Attention must be given constantly to see that the crew members catch the samples at the proper intervals, and that they do not anticipate the sample by filling several sacks at one time.

In pay sections where the exact measurements are very important, or in deep holes where the cuttings take a long time to come to the surface, samples should be labeled with the depth at which they are actually cut rather than the depth of the well at the time they come to the surface.

The mechanics of depth control are varied, but there are several methods in use to assure accuracy.

There are many ways to catch representative samples from the returning mud stream. The rotary shale shaker, the vibratory shaker, and the sample trap are some that are in common use. After they are caught the samples must be washed properly, then dried. The drying is generally done artificially, using care to not overheat them so that natural oil staining will not be blackened and mask porosity.

After the samples have been caught and processed properly, they are ready to be examined by the geologist. There are two principal ways of describing samples. The first of these is the interpretative system, in which the geologist picks out the cuttings which he believes to be representative of the formation penetrated and describes the entire sample as composed of this rock. The rest is assumed to be cavings. This kind of description brings out formation changes and is of greatest value in areas where the various formations are of wide extent and relative constant character. In regions of pronounced lateral gradation it has been found that a second method is most satisfactory. This is the percentage description, where the geologist describes all the material in the sample, disregarding obvious foreign substances and cavings. This system, though making it difficult to determine formational boundaries, shows the gradations of the beds and often enables one to trace a horizon through different sedimentary facies.

Once the description has been made, it must be plotted in graphic form to make the information easily available for correlation and study. It is usual to plot this material on a narrow strip of heavy paper or cardboard, so that a large number of logs can be laid out to compare and correlate the sections.

Uses of the sample logs include stratigraphic correlation for structural mapping, analysis and evaluation of any pay zones, a study of the characteristics of complex reservoirs as a supplement to any coring program undertaken as part of a reservoir study, and selection of points for plug-back or for packer settings for remedial work.

§ 6.03 Mud Logs

The determination of the fluid content of the porous formations penetrated by the bit so that no oil or gas bearing formations will be overlooked, is one of the most important problems in drilling for oil or gas. Modern mud analysis, well logging equipment, and improved techniques allow for this detection of very small quantities of oil or gas and the placing of these shows at the proper depth.

The basis for mud logging is that when the well is drilled the bit disintegrates a cylindrical section of the formation. If the pore spaces of this cylinder contain oil or gas, part of these contents will be contrained in the drilling mud and part will be retained in or on the cuttings. If the drilling fluid and the cuttings are continuously tested on their return to the surface, and the results of these tests correlated with the depth of origin, the presence or absence of oil or gas in specific formations can be determined.

The test for gas in the mud is made by diverting a portion of the circulating mud from the flow line to a separator or gas trap, where the mud is removed. Air is mixed with the gas and this air-gas mixture is then drawn into a "hot wire" gas detector instrument, where the percentage of combustible gas is determined. The amount of methane gas present in the total is determined by controlling the temperature of the filament in the gas detector.

The gas readings in the mud are recorded as total gas (which includes all the combustible gases) and methane gas. Some methods also use a chromatograph which determines the presence of ethane, propane, butane, and pentane. This "breakdown" of gas readings is made because studies have indicated that, with few exceptions, all productive formations logged by the mud analysis method have shown a definite increase of methane gas. Normally those zones that do not show an increase of methane may be condemned.

The test for gas in the cuttings is made by placing a small sample of the cuttings with a measured amount of water in a closed-container-type high speed grinder. After cuttings are ground, the air-gas mixture in the container is examined for gas in the same manner as described before. Here, too, the gas is reported as total gas and methane gas.

The presence of oil in the drilling fluid is detected by a physical examination of the drilling mud under ultraviolet light. A sample of the mud is treated to reduce the surface tension and gel strength, after which it is placed in a viewing box. This box is constructed so all external light is excluded, thus the sample may be selectively subjected to either ultraviolet or white light. The magnitude of the oil shows in fluid logging units is based on the amount of observed florescence of the crude oil.

A depth meter, pump-stroke counter, and pump-rate meter provide the necessary data for obtaining a rate-of-penetration curve, and for properly correlating with depth any shows of oil or gas found in the mud and cuttings at the surface. The log is prepared as a plot of the magnitude of the oil and gas shows versus depth. Also presented is a lithology column made from the cuttings, as well as a description of the material that makes up the cuttings.

The mud logging service is used as wells are being drilled, and results on any formation are available as soon as the mud used in drilling the formation reaches the surface. The system does not interfere with the drilling operations in any way, and no changes in ordinary drilling equipment or procedures are required to make it workable. The logging units are mobile field laboratories. In addition to the logging instruments, each unit is equipped with complete mud testing equipment so that mud weight, viscosity, gel strength, water loss, cake thickness, salinity, and other characteristics may be maintained as required.

The principal application has been in the drilling of wildcat or exploratory wells, where its usefulness is apparent in many phases of the well program. Coring can be reduced to a minimum by being able to choose only those sections which contain oil or gas. This type of logging can be depended on to pick up indications of all possible productive formations.

The method does not present a complete subsurface picture, however. It does not give quantitative determinations of the amount of oil and gas occuring in the formation, nor does it furnish quantitative information on the productivity of the oil or gas formations. However, the method does give reliable qualitative information on the occurrence of oil and gas, and the interpretation should be made in the light of all available information on that section of the hole logged. It is extremely valuable in minimizing coring, in removing the ever-present possibility that a producing zone will be missed, and in rounding out the picture presented by other formation data.

Figure 85 shows a typical mud analysis log. Column 1 is the drilling rate in minutes per foot, Column 2 is the visual porosity in the cuttings, Column 3 is the lithology from the cuttings, Column 4 is the depth column, Column 5 is the chromatograph gas breakdown, Column 6 is florescence in the mud, Column 7 is total gas and methane, and Column 8 is the remarks column where formation description, mud data, and drill stem tests are listed.

Figure 86 is a mud analysis log with the electrical well log attached for correlation. Column 1 is the Dual Induction Log, Column 2 is the drilling time in minutes per foot, Column 3 is the visual porosity of the cuttings, Column 4 is the depth column, Column 5 is the lithology from the cuttings, Column 6 is the oil in the mud, Column 7 is the florescence in the mud, Column 8 is the total gas in the mud, Column 9 is the chromatograph breakdown of the gas in the mud, and Column 10 is the formation description, mud data and drillstem test column.

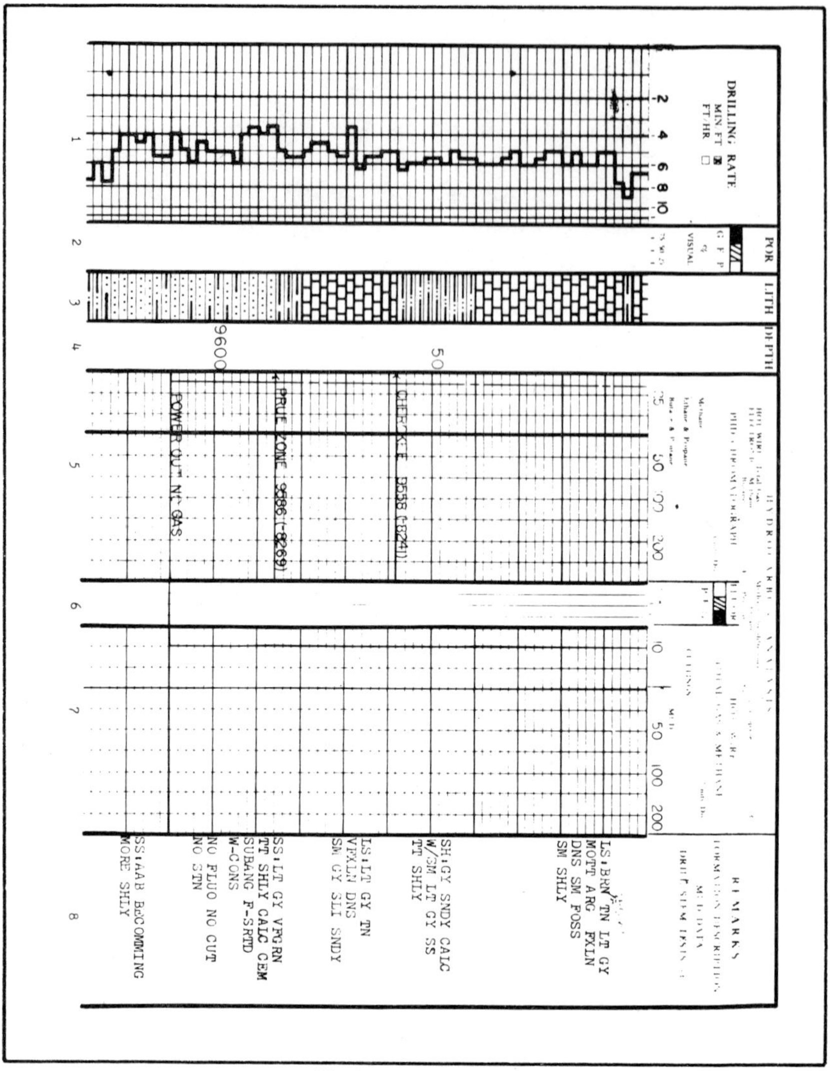

Mud Analysis Log
Figure 85

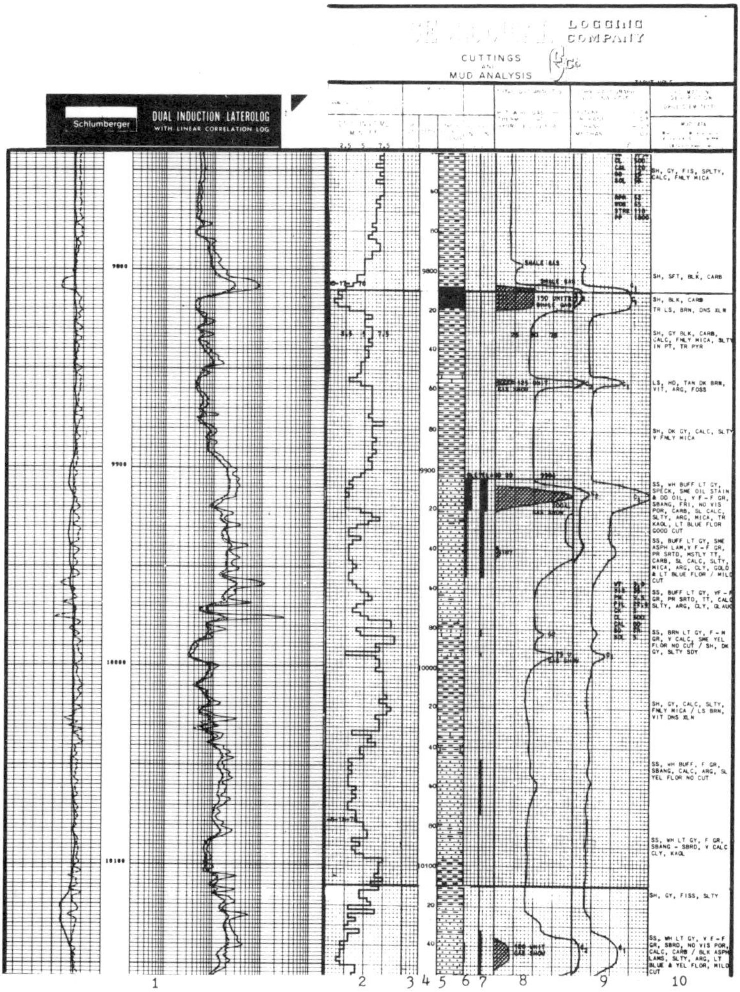

Mud Analysis Log With Dual Induction Log
Figure 86

Chapter Seven

REFERENCES AND CREDITS

There are great quantities of publications available pertaining to well logging. There are those that describe the theory of every measurement discussed in this handbook, there are some that describe the instruments necessary to make these measurements, and then, of course, there are many that delve into the analysis and interpretation techniques.

§ 7.01 Authors

The authors who research and prepare these papers come from all segments of the industry, ranging from service companies and oil companies to schools and consultants. They are presented at professional society conventions, and in industry publications. The purpose is to introduce new services and explain new techniques and procedures, and to complement existing techniques and procedures. New publications are constantly being presented.

Some of the companies that have publications available include: Dresser - Atlas, Gearhart Owens Industries, Schlumberger Well Services, and Welex. They each have "Technical Bulletins" which explain in detail every individual logging method and technique, as well as books of "Log Interpretation Charts" which provide charts used in applying the various interpretation and analysis techniques. Schlumberger also has available interpretation "Documents" which cover particular subjects.

The paper referred to concerning Figure 84 is titled "Using Log Derived Values of Water Saturation and Porosity" by R. L. Morris and W. P. Biggs.

Chapter 6 contains excerpts from a training textbook prepared by this author for Core Laboratories, Inc. in 1973.

The author expresses his appreciation to Kenny Kemp with Schlumberger and Bob Parmley with Geological Logging Company for their help in procurring some of the illustrations.

INDEX

E

F

G

H

I

L

M

N

P

Q

R

S

T

U

V

W